T0273783

# The Construction Technology Handbook

# The Construction Technology Handbook

**HUGH SEATON**

WILEY

Published by John Wiley & Sons, Inc., Hoboken, New Jersey.
Published simultaneously in Canada.

For general information on our other products and services or for technical support, please contact our Customer Care Department within the United States at (800) 762-2974, outside the United States at (317) 572-3993, or fax (317) 572-4002.
Wiley publishes in a variety of print and electronic formats and by print-on-demand. Some material included with standard print versions of this book may not be included in e-books or in print-on-demand. If this book refers to media such as a CD or DVD that is not included in the version you purchased, you may download this material at http://booksupport.wiley.com. For more information about Wiley products, visit www.wiley.com.

*Library of Congress Cataloging-in-Publication Data*

Names: Seaton, Hugh (Tech entrepreneur), author. | John Wiley & Sons, Ltd.,
    publisher.
Title: The construction technology handbook / Hugh Seaton.
Description: Hoboken, New Jersey : Wiley, [2021] | Includes bibliographical
    references and index.
Identifiers: LCCN 2020038949 (print) | LCCN 2020038950 (ebook) | ISBN
    9781119719953 (cloth) | ISBN 9781119719908 (adobe pdf) | ISBN
    9781119719977 (epub)
Subjects: LCSH: Building—Data processing. | Construction
    industry—Technological innovations. | Information technology.
Classification: LCC TH437 .S43 2021 (print) | LCC TH437 (ebook) | DDC
    624.028/5—dc23
LC record available at https://lccn.loc.gov/2020038949
LC ebook record available at https://lccn.loc.gov/2020038950

Cover Design: Wiley
Cover Image: © matrioshka/Shutterstock

SKY10022609_111820

*This book is dedicated to my mother, Bonnie Verses.*
*Thank you for being there.*

*And to all the men and women in construction,*
*the greatest industry in the world.*

# Contents

# Foreword

"We must keep up or be left behind ... "

When Hugh and I first met a few years ago, I almost immediately mentioned this phrase. Hugh, being an advocate of virtual reality (VR)/artificial intelligence (AI) and all things technology, wanted to deep dive into our world of construction and see what was there. We began discussing ways we could integrate VR and AI into our current training workflow. Hugh was very enthusiastic about furthering the development of training in the trades.

First, Hugh wanted to know our thought process on construction technology and where our motivation was. We wanted to start from the beginning, interested in the way it was done in the early days when mobile devices and 3D-generated models were not the mainstay. We decided that we needed to have some conversations with those who spent a great deal of time in the field.

We pulled some of our *tenured* members into these conversations. Our brothers and sisters who have spent the last 40 years in the trades, on the verge of retirement. We asked them about the modern jobsite and how it was different from "way back when." We asked what their opinions on technology were then and if their opinions are different today. We had conversations about productivity and efficiency. We were curious as to how technology could be an impact on the momentum of a jobsite.

We spent a great deal of time breaking down these conversations, attempting to tie it all together. Looking for the breaking point, the moment when technology took over and became commonplace. We wanted to put our fingers on the exact date. What we kept circling back to was frustration. Let me explain.

"So, how do you feel about your company's VDC department?"
*"What's a VDC department?"*
"Not a big deal, how about BIM ... at what point do you feel BIM took over on the jobsite?
*"What's BIM?"*
"Great question. Let me ask you about iPads. Would you rather have paper drawings or an iPad in the field?"

---

*"Paper drawings, no questions asked."*

---

Thinking that we would have the answers put in the palms of our hands, we were quickly awoken to the fact that technology had come on so quickly most of those in the field didn't have an opportunity to fully grasp it. A whirlwind of change came, and they were swept up with it. They were frustrated. They wanted to go back to how it was done in the past. Nobody spent the necessary time with them explaining the benefits of these new tools.

They were not given the proper training, they had no idea what these new TLAs – *three letter acronyms* – meant. It was assumed that they would be able to keep up. Most importantly, their feedback was never a conversation point. We never really asked them – the true professionals who spent the last 40 years in the field – how to properly integrate these new technologies. We failed to ask them how they would strategically integrate a new tool into the daily workflow.

We sat back, expressionless, realizing we were going about it the wrong way for a long time.

This book is intended for anyone who lives on the modern jobsite. Whether you are new to the construction industry or have years of experience behind you, this book will break down technology in an easy-to-read format. It will give you the resources you need to have conversations on the jobsite about technology.

This book will empower you to innovate and change the way things are done. In order to succeed we must all have a voice and we must all pitch in.

*If we do not keep up, we will be left behind.*

–Mike Zivanovic

# Preface

This book was written to bridge a gap between the technology world and the construction industry. It provides a collection of definitions, explanations, and discussions about everything from what technology is, to how it works, to how to innovate.

Technology is just another set of tools, and these are supposed to be easier to use than older tools. Some are not, but most are trying to become easy, fast, and useful. By understanding the terms and some of the concepts, you will find new technologies easier to try out and master.

Most of all, I want to dispel the myth that somehow technology is "different" from the work that goes on in construction. Everyone uses technology every day. Digital technology takes a little translation, and a little getting used to, but it is not even close to the hardest thing a pipefitter, mechanical contractor, surveyor, or any one of the seasoned professionals onsite or in the office need to know to pull modern buildings out of the ground.

Human intelligence, problem-solving ability, and general common sense are irreplaceable. No software, robot, or artificial intelligence that we can build or even conceive of can do what construction professionals do every day.

Reading this book will add to your toolkit, so you can go out and build the world faster, safer, and, hopefully, a little better.

In addition, because any book will get outdated almost immediately, I will be producing a quarterly round up of construction technology, *The Construction Technology Quarterly*. It will comprise a free, downloadable report, and a free presentation webinar. You can learn more at https://www.constructiontechnologyquarterly.com/

Hugh Seaton
New York, NY
June 2020

# Acknowledgments

I set out to write a book that would be useful to real people, so I asked as many people as I could find. The construction industry is full of down-to-earth, smart people who like the idea of sharing their thinking and in the process, molding mine. It is a bigger list than most books, and my debt to the industry is greater than most authors. I am humbled by your insights, and honored to have heard you.

Paul Doherty introduced me to BIM in 2010 and has been a friend and mentor ever since. Damon Hernandez introduced me to Silicon Valley, Virtual and Augmented Reality, and showed me how to run a hackathon – his friendship has been a defining influence for almost a decade. Cody Nowak, also of hackathon fame, took to my writing of this book like a true friend and introduced me to dozens of his colleagues. Mike Zivanovic has been a guide to the trades and the ultimate gut-check.

These four were instrumental to the success of this book – thank you guys.

Thanks to Al Vaquez, I really understand what a world-class software engineer can do, and thanks to the Glimpse Group, I've seen what a committed, smart group of technologists can do. Thank you especially to Lyron Bentovim, Maydan Rothblum, and Saul Pena.

To Sasha Reed and Jessie Davidson of Procore.org, thank you for the opportunity to create the "Data in Construction" courses, and for putting up with me while I finished the book.

I was lucky enough to interview dozens of people for this book, all of whom contributed to my understanding, all of whom tried their best to keep me out of trouble. I'm honored all of you would spend the time to share your wisdom. Thank you to:

Aarni Heiskanen of AEC-Business.com, for telling me about construction technology in Europe

Abhya Sinha of DPR, for telling me about data and VDC

Alex Brown of Openspace.ai, for a great intro to capturing jobsite progress

Amy Marks of Autodesk, for schooling me on Industrialized Construction

Andy Huh of Fentrend and SCS-NY, for insights on startups in construction

Atul Khanzode of DPR, for breakthrough thinking and enduring a "fan-boy" interview

Barry LePatner, founder of LePatner & Associates, for clarifying the muddy waters of construction contracts

Blake Berg, chapter lead of the SCS-NY, for insight into tech in the field

Brek Goin of Hammr, for insight into the trades

Cherise Lakeside of CSI, for amazing insight into the demographics of the industry

Chris Tisdel of Ruckus Consulting, for telling me about technology in construction

CJ Best of McKinstry, for some great cases of data and technology in the trades

Dan Bulley of the MCA Chicago, for amazing perspective

Dan Nash of Kiewit, for sharing perspective on innovation at GCs

Danielle Dy Buncio of ViaTechnik, for perspectives on technology in this complex industry

Darren Young of Hermanson, for perspectives of a construction technologist

David de Yarza of BuilderBox, for entertaining perspectives on contracts and innovation in construction

Don Metcalf of Nemmer Electric, for a real-world view of prefab and offsite construction

Doug Chambers of Fieldlens and WeWork, for support and perspective on startups in construction

Hamzah Shanbari of The Haskell Company, for insight into how they do innovation

Harry Handorf of Holobuilder, for explaining the future of construction site imaging

Heather Wilshart-Smith of Jacobs, for insights into data in construction

Jake Olsen of Dado, for amazing perspective on how to create technology people actually need

James Benham of JBKnowledge, for great perspectives and putting in the years to tranform the industry

Jamie Frankel of Schiff Hardin, for guidance as I researched the book

Jeff Sample of eSub, for perspectives on selling and supporting software in construction

Jesse Devitte of Borealis and Building Ventures, for a revealing look at the past and present of construction technology startups

Jonathan Marsh of Steeltoe Consulting, for a brass-tacks look at technology in the mechanical trades

Josh Bone of NECA, for being the coolest guy in construction, and generously sharing his time for my various projects

Karl Sorenson of Blue Collar Capital Partners, for early encouragement and great perspectives

Kaustubh Pandya of Brick & Mortar, for bringing high-level VC perspectives

Kean Walmsley of Autodesk, for the future of design perspective

Ken Schneider of the United Association, for support and perspective

Ken Simonson of the Association of General Contractors, for an economist's perspective

KP Reddy of Shadow Ventures, for perspectives on startups

Kris Lengieza of Procore, for some great cases of data and technology in construction

Marc Kinsman of Mortensen, for insight into how VR can help GCs

Marco Faccini, for an English perspective

Martyn Day, for a pointed English perspective, especially on design tools

Matt Carli of Latticrete, for insight into the technology of materials

Matt Daly of Structionsite, for insights into technology on the jobsite

Matt Diesner of Autodesk, for a perspective on sales in construction

Mike Prefling, for sharing stories of innovation in construction

Mostafa Akbari-Hochberg of Holobuilder, for explaining the future of construction site imaging

Nathan Wood of the Construction Progress Coalition, for inspiration and deep insights

Ned Beatty of IrisVR, for thoughts on virtual reality in construction

Pat Sharpe of The Digit Group, for being a friend and endless source of insight

Quinn Murphy of Sandberg Phoenix, for telling me technology brings transparency, which is a good thing

Ricardo Khan of Mortensen, for pushing the industry forward and showing us what innovation looks like

Richard Harpham of Katerra, for a blindingly insightful first talk that showed me how big these issues are

Rob Fischer of CURT, for an owner's perspective and some great cases of how owners can drive everything

Robert Friedman of TechPrefab, for an excellent deep dive into Prefab

Sam Spata of Exyte, for a great explanation of Lean Construction

Shane Scranton of IrisVR, for thoughts on virtual reality in construction

Stefan Larsson of BIMObject, for a vision of what BIM could be

Steve Holzer of BIMObject, for specific examples of what BIM *should* be

Steve Jones of Dodge Analytics, for a great overview of data in the industry

Tauhira Ali of Milwaukee Tool, for helping me understand how software is reinventing hardware

Taylor Cupp of Mortensen, for great perspectives of a construction technologist

Teemu Lehtinen of KIRA Hub in Finland, for a perspective on Finnish innovation

Terry Cotton of SAM Floors, for a supply chain perspective

Tim Etherington of Gensler, for a truly global perspective on architecture, from China to Spain and back

Tim Hensley of Hensel Phelps, for a patient walkthrough of how a Senior Superintendent uses tech on the jobsite

Todd Mustard of TUAC, for perspectives on associations as drivers of innovations

Tony Bruno of Omnibuild, for explaining how he uses construction tech on the jobsite

Travis Voss, for a vision of what a rockstar technologist can bring to their company

These folks and more have done their best to help me see what's going on – any failure to get it right is my own, not theirs.

# CHAPTER 1

# Introduction

How you think about the world affects what you can get done in the world.

By thinking differently, you can do different things. Books like this one expand how you think, and will therefore expand what you are able to do – not because of quickly outdated "how to" lessons, but because of powerful frameworks for viewing all of what you do as a kind of technology, and viewing new technologies not as separate from what you do, but simply new tools in an expanded toolkit.

This is a book about technology that is used in construction. "Technology" is one of those words that gets used differently by different people, which makes it hard to discuss. To be able to think clearly, differently, we need a concrete definition of what words like "technology" mean. In fact, the first point I want you to agree with, accept, and internalize is that you cannot think clearly with fuzzy concepts, and technology will introduce you to a lot of concepts that are fuzzy to you at first. In this book, we will stop and define as many new terms as possible.

Construction is an industry composed of trades and practices that are taught as much by showing as by talking, so the culture isn't always one of directly asking people that you don't know what they are talking about. There can be a sense of discomfort about asking, because at some point technology, especially software, has made everyone feel stupid.

Read this book and that will happen less, I promise. However, the point is to feel confident that it's not your ignorance of whatever new concept is being discussed, but the vendor's or presenter's failure to make sure there is common understanding.

In the case of technology products and processes, it is *always* the job of the provider to make sure you are clear – hold them to it.

## What Technology Is

So, let's get in that habit of clear definitions by creating one for technology:

**Technology is the application of some effect, usually scientific, to get work done.**

The word "technology" can also be used for two other levels of meaning:

1. A collection of things that work similarly, like *construction technology*.
2. The whole class of human effort that creates tools for a given culture, like *digital technology*.

We are going to focus on the first meaning. It is important to think at this level first, because you will be dealing with specific products not big groupings or abstract classes of products.

When faced with a new technological product, like construction software, we can be struck by what we don't know, struck by how different it feels from how we've done things in the past. But technologies do not come from nowhere. To be of any use, a new machine, process, or software will have been developed so you can do something you already do, just faster, safer, or cheaper.

### Understanding a Technology's Basis

Technology of any sort is based on some underlying effect, some realization that nature, or human nature, works a certain way. There is some effect, or phenomenon, that makes the technology work. So we build a process, or a tool, or a machine, that exploits this effect to make human work better in some way. Often, these technologies make impossible things possible.

For example, think of a hammer. We don't think of this as a technology, but it is. Here are some of the effects in the world that a modern, handheld hammer exploits:

1. Every force creates an equal and opposite force (Newton's third law, the same one used in rockets)
2. Steel is hard
3. Cold rolled, high carbon steel is very hard
4. Metal is harder than wood or gypsum
5. The end of a pendulum is faster than the handle
6. Force applied to a given area gets multiplied when transferred to a smaller area

All of that in a simple hammer. Think then of what a hammer does: it uses motion from a human arm to transfer force from one steel object, the hammer's head, into another steel object, the nail. This force then drives the nail through whatever material is being worked on.

Let's take a moment and think about what you do, all day long. Whether it's putting electrical conduits in place, managing a team of mechanical contractors, managing a jobsite as a superintendent, or managing an entire job as the project manager – *everything* you do works because of some effect in the world. Some of those effects are very human, like ego, pride, and a desire to create something real in the world. And you learn through your career to use those effects to motivate, manage, or just navigate other people. For example, you learn to check up on people frequently because you know that accountability makes people more focused on the job – an effect you leverage to get the job done.

Managing is a technology every bit as complex as artificial intelligence – in fact, as someone who has done both I can tell you managing can be harder because it is a never-ending balancing act. Management as a practice has evolved over time to use different methods, each using a different effect in the world – we used to rely solely on hierarchy and power, which relied on a fear of losing one's job. But we realized that stifles critical information flows and causes worker disengagement. So we've swapped the underlying effect to one of a feeling of involvement and achievement, which is what Lean Construction is focused on. Changing the effect a technology is based on can be very powerful.

Back to our hammer example, what happens if we separate the work to be done, driving the nail into the wall, from how it gets done, the centrifugal force of a swung piece of steel hitting a stationary nail head? What if instead we put the nail in a tube connected to compressed air, and "shoot" the nail into the wall?

We've changed the effect being exploited to get the same work done, from a human arm swing to mechanical air pressure release, and in the process have dramatically improved the efficiency of our nail-driving workers, significantly improved their ability to keep driving nails without fatigue or arm injury, and hopefully spared their thumbnails.

## Technology Domains

When we think of technology as a chain of these effects being used, we can more easily understand how to integrate new pieces. Technology is just a tool that is an extension of human power and action, no more, no less. By keeping in mind that any technology is there to extend *your* power and action, you can put it into a context that helps both assess how good that technology is, and understand why new technologies might be better than old ones – they are using a newer, better effect to help you. In fact, it is more often the case that the tools you use are exploiting a bundle of effects to work, and the part you will care most about, the effects being exploited that matter to you most are human and organizational effects.

And by understanding all technology as being part of a chain you are part of, you can be clearer with the creators of that technology about how they should develop and deliver real products. What is happening in construction now is that parts of the chain of technology, starting with your skills and extending to the tools you use, are being added and changed. Understanding modern technology not as a separate class of things, but just the latest in a series of "modules" that can be swapped into an existing chain is a much better perspective, because it keeps technology under the category of "tools" – not something alien.

Technologies themselves are always a collection of other technologies. Even something as simple as a hammer is the result of plastics, metallurgy, metal working, factory design, automation, ergonomics, and even packaging. Knowing that technologies are themselves composed of interlocking, lower level technologies, makes it easier to understand how our tools keep getting a little better, cheaper, and safer every year. For example, Milwaukee Tool has been making tools like torque wrenches for years, and over time they have swapped out some controls that were mechanical, like a direct lever between the trigger and the electrical motor, with electronics that give the user more operational fine tuning.

Milwaukee has recently added a new feature, their "One-Key System," that wirelessly keeps track of all the tools on a jobsite. You can look at this as just another feature, actually a pretty cool one, or you can look at it as Milwaukee swapping out the paper and pencil, or perhaps spreadsheet technology that relied on observation, memory, and making the rounds to keep up to date, with an automatic, real-time inventory reporting tool. Your chain of technology has not changed, but the tool you use to do a part of it has.

Thinking of your job as a chain of tools, from your own skills to the external tools you use gives you, the professional using these technologies, the power to think creatively about what you're using, combine products and technologies in new ways, and demand that providers of technology products keep working until the products do what you need them to.

This idea of changing out one kind of technology for another is often referred to as changing the "domain" of technology being used. In the hammer example, we changed a manual force domain for a pneumatic force one. In the Milwaukee tools example, we changed from a manual inspection and reporting domain to a wireless, automatic updating domain.

This "redomaining" or swapping of one technology domain for another in human activity has been happening since the dawn of time. These new domains tend to come in waves, and as they do, there is an inevitable process of blending the seasoning and judgment of industry professionals with the new processes and skills that these new technologies bring with them. That's what we're in the middle of as an industry, replacing physical and industrial-era techniques and technologies with digital domains.

From the examples above, human work was replaced by machines. Fears abound in the specialty trades, and contractors more generally, that technology is going to cause people to lose their jobs. We will address this very specifically in the AI and Industrialized Construction chapters, but for now, keep two points in mind: the first is that machines and software replace skills from the very bottom, starting with the most mindless tasks. Some of those tasks still require skill, like using a hammer well, but those skills are a small part of what makes a carpenter a professional. The second point is that what does make a carpenter, or pipefitter, or mechanical contractor a professional is the ability to solve problems, and to come up with solutions to complex obstacles that involve schedules, contractual terms, team dynamics, and many other things.

## Domains Versus Products

It can seem that changes in technology are inevitable, and when looked at as a collection of solutions and products, there is some degree of truth in that. Science will keep producing new effects and insights we can leverage. Companies, driven by the designer to out-compete each other, will keep refining ways to exploit those effects.

But there is nothing inevitable about any given technology. We in the industry can absolutely affect what products are out there, and more importantly, how they work for us, and with us.

An example of this contrast between a kind of technology that was probably going to happen no matter what, and specific products that were very much not inevitable is the VHS/Beta battle in the 1980s.

Starting in the 1970s, film and TV were revolutionized by digital storage, which became video cassettes. There were two product options: Betamax and VHS. Beta was higher quality video, because its developer, Sony, assumed consumers would want to enjoy their movies and TV at the highest possible visual quality, which they achieved by limiting the playback to one hour. Panasonic's VHS, in contrast, provided lower visual quality but up to 2 hours in length when first introduced. Since most movies are over an hour, Beta didn't fit the market as well, and ended up losing the consumer market to VHS, which became the standard for about a decade.

Two things were at play – a standard and a product. In this case the consumer videocassette market became dominated by the VHS standard, and specific products were all VHS.

Prior to this, the only way to view a movie at home was when a broadcast network chose to air the movie, which they did rarely and in highly edited, kid-friendly form. The home video market was thus "redomained" so that the way consumers viewed movies at home went from a broadcast-centered model, to a videocassette-centered model. The new domain was going to happen, but the specific products would win or lose based on how well they fit the market. And you, dear reader, are the market. You decide who wins or loses.

This matters more than it might seem, because when a new domain emerges, there are often a ton of options for a little while. Some of them fail, some get bought or merged with others, and a few will win. Everyone knows about Facebook, and some might remember MySpace. But do you remember Friendster and the 40 or so other social networks that came out in the mid-2000s?

That's happening right now with construction project management software, where beginning in about 2015, more and more companies have set about digitizing the construction workflow – changing the technology domain from paper and Microsoft Excel, to unified platforms that deal with different parts of the construction process, or all of it. In time, there will be winners and losers – probably not the monopolies we see in consumer markets, but definitely fewer product offerings.

Redomaining can often come from other industries. Take building security, for example. The videocassette made possible an entirely new capability for the capture and storage of video from cameras in a building, cameras which themselves had been changed from film to digital in around the same 1980s' timeframe.

But in the later 1990s, content of all kinds, from music to video and images all began to convert into digital formats. These started as CDs, but then changed into MPEG video and MP3 audio. Around 2000, consumer markets started marketing players for these digital formats, which drove down the cost for digital storage for security, leading to today's systems that are entirely digital. In fact, systems like building security have seen a series of technological domain changes, from videocassettes, to CDs, to hard drives, to the cloud most recently.

It is often the case that developments in other markets, especially consumer markets, create pressure to change in other markets, because the technology becomes cheap and familiar to users.

These model changes, the redomaining, is going to happen no matter what. But the specific *way* it happens is not inevitable. This is an important lesson for technology in construction: pressures of technological advance mean that the construction process will continue to digitize, will continue to absorb and integrate new technologies, but any given company or product could succeed or fail. How to assess these products will be one of the key takeaways of this book, as we go through each of the technology areas that you will encounter.

## What's Different about Digital Technology

No one needed to write a Construction Technology Handbook when the technologies being used were confined to the individual trades, and involved tools and machinery that exploited mostly physical effects. Learning to use a nailgun was not a huge leap from a hammer; learning to use increasingly powerful and sophisticated power tools was usually an evolutionary process where features kept getting better.

In these instances, workers can *see* how the technology works, can understand intuitively how to at least use the technology, even if it might take years to master the craft overall.

In contrast, digital technology does its work out of sight, in non-physical ways that humans cannot immediately grasp, using controls that aren't "natural" in the way a hammer's handle is. Older technologies exploit physics, which humans naturally understand.

Digital technologies exploit electronic phenomena, which are so small that we cannot see them. Digital technologies also build layers of human-designed interfaces that don't have to rely on intuition or natural movement at all, they are completely the invention of the product developer. And that means the intuition and experience that work so well with physics-based technologies don't help us with digital technologies, and that can be alienating and annoying. These human-designed interfaces do have logic, though, and are based on real engineering – so you can learn that logic and become just as comfortable with digital technologies as you are with anything else.

That logic, that "physics" of digital technology is what this book is about. We have already started from the beginning, defining what we mean, and will build up the rules and frameworks that give you a deep understanding of what makes digital technologies tick.

## Your Community

For years, the Architecture, Engineering and Construction (AEC) industry has been among the slowest to adopt technology, in part because the technology that was introduced early on was often not well adapted to the needs of the industry and its different parts. In the past decade or so, a number of community efforts have arisen to give the AEC industry, especially construction professionals, opportunities to learn about technology, while giving the technology industry a window into one of the oldest, most complex industries.

Perhaps the first of these was the AEC Hackathon, an event series that I helped kick off in 2013. Founded by Damon Hernandez and Paul Doherty, the event was created to break down barriers between AEC industry people and tech people. By solving real problems together, both sides get an appreciation for the other. We've run over 50 of these around the world, changing the format to an online version in the post-pandemic world. Now that they are digital, I encourage you to check one out; go to www.hackaec.com to see what's out there, and find others like you who are on a digital journey.

In the years since 2013, startups and venture capital have discovered construction, and there has been a flood of solutions for everything from 3D scanning to daily reports. Not all of these came from a good understanding of what's really needed on a jobsite, and in fact going back a little further even more of the software pushed to the construction site came from other places, like accounting. We've heard stories of "app-fatigue," and a general concern that field personnel especially are not a big enough part of the tech development and adoption process.

We need to change that.

In this book, I share new skills, and a new mindset toward technology. Whether you're already a construction technologist and have "drunk the kool-aid," or worried about how technology will impact your life, you will find ideas of value in these pages.

Through this new digital toolkit you'll look at technology differently, from the inside of how it gets made to how it gets packaged and adopted. You'll also see that the people who make software and other kinds of technology are very similar to construction, especially the trades. The big difference is where construction expertise is aimed at putting work in place, technology expertise is aimed at "making it go." Both sets of professionals take enormous pride in what they do, and by understanding technology as just another toolkit, you can blend the strengths of both the construction and the technology mindsets.

It is often said that training teaches you how to do, education teaches you how to think. A book like this is educational – so ask questions along the way, and try to see things with the mindset of a technologist.

## Mindset Matters

What do we mean by "mindset?"

We started the book with an assertion: How you think about the world changes what you do, and how you do it.

That is a mindset. And the promise that idea makes is this: Change your mindset and you can change your possibilities. Anyone who's played sports will agree that mindset is everything.

What specifically does a changed mindset change in your real work or life? For a start, it changes what you pay attention to, and what you think is worth your time.

In a complex, fast-moving environment like construction, you have to pick out what matters from a sea of events, meetings, and messages. You can't figure out what matters if you don't have an idea about how the world works – a model of what causes what, and what is important in the end.

Construction has always had a huge toolkit, from MIG welders to hand tools to heavy machinery, and everything in between. Most of those tools are based on processes that have evolved slowly and, as discussed above, are based on observable effects. To make those processes successful, the construction mindset has been one of relying on past experience, trusting your gut, and a constant anxiety about what might go wrong that you cannot see. This mindset is why superintendents, in fact why almost everyone, is constantly inspecting the jobsite – they are *looking* directly at everything, only trusting their own eyes.

The mindset and mental models that made one good at construction traditionally are not the same as the mindset that will make you a master of both the trade you already know, and the digital tools that you use now and will be able to use in the future. There is no reason you cannot use both mindsets, because they are not really in conflict.

The digital construction mindset separates the physical from the digital, understanding that each supports and requires the other to function. Understand that some problems are for the gut, and some are for the analytics. These two higher-level toolkits are critical, and they also overlap.

## Intuitive Problems

Processes and problems that you can directly see are always going to be where experience and intuition are primary. No machine or software can compare to the sophistication that a seasoned professional brings to a real situation.

This is true for a few reasons, the most important of which is that no software has been invented, including the much hyped artificial intelligence (AI), that has an understanding of context. Humans are highly context-driven, and we understand not just what's right in front of us, but also how it relates to what happened yesterday, how it might relate to the specific men and women on the job, and other factors. Digital technology cannot do this.

But context doesn't only come from what you've heard or seen, it can also come from digital technology – reports and analyses of a bigger picture, that help your intuition do its job.

## Digital Problems

Intuition works by simplifying the world, by mental rules of thumb, so that we can take all the information in and do something quickly. We evolved to be fast on our feet, to see issues immediately, which is why we are so good at managing what we can see.

However, modern construction involves huge numbers of people, across months of work and multiple sites. That is too much for our brains to handle directly, so we either use reams of paper and tons of meetings to keep it all straight, or we change the domain of technology we use to digital software.

Digital problems, involve lots of data, either because of a large scope of information, or really fine measurements only a machine can make.

For example, tracking events and progress across a company, or even a job, will include thousands of points of data, over weeks and months of a project's lifetime. This is not what our brains are good at, and this is where digital technology helps – with what we cannot directly experience with our senses.

The analyses that come from digital software can show us patterns that we would not otherwise see, and at least as importantly, create *proof* that these patterns exist, so we can discuss them with management and start to create improvements.

But just as analytics can provide context for intuition, intuitive "gut-checks" on data and software are absolutely critical. There is no software in existence that can really correct itself when it has bad data or other unexpected issues. Human oversight of software, robots, and other digital technology is essential.

Our mindset shift, then, is to accept digital technology as a tool for seeing the scale and scope of information that cannot be directly seen. We need to build trust that these technologies can do what humans alone cannot, but also understand their inherent limits.

We'll develop a clear understanding of what technologies can, and cannot, augment how you already do things. Just as today's manufacturing includes all of the experience, intuition, and "gut" from experienced managers and applies it to incredibly advanced machines, software, and analytics, technology as a part of your construction toolkit only makes sense if it adds to what you are already good at.

Nothing can replace the almost magical ability of the human mind to understand what's going on and with deceptive ease, know what to do next. This is a book about technology, but the most important technology of all is how you think.

This construction technology mindset will change how we build the world. Let's see how, with an example from manufacturing, specifically the Lean mindset.

## How the Lean Mindset Changed Manufacturing

Car manufacturing in Japan is actually where all of "Lean" came from – based on the innovations of Kiichiro Toyoda and Taiichi Ohno, the bosses of Toyota motor company in the years after World War II. Faced with almost no capital in a capital-intensive industry, workers with limited skills in a skills-intensive industry, and almost unbeatable competition from Detroit in a competitive market, Toyoda and Ohno had to find a new way of looking at their business.

They did this by going back to first principles, and questioning everything about how cars were manufactured, to see how they might invent a new way. The first thing they questioned was how success was measured on the factory floor – traditionally this had been driven by the fact that the big pressing and assembly line machines were (and still are), enormously expensive. So the goal from a financial standpoint was always to make sure you maximize return on investment (ROI).

Following this ROI mindset, traditional car company managers and floor operators worked to make sure that each machine was used as much as possible, which meant producing as many parts produced per hour as possible, so the cost of the machine could be spread out across those parts. In this ROI-centered mindset, managers valued machine efficiency more than the efficiency of the whole process.

Think about what that means if you have, say, five machines in a row making a screwdriver. The first one extrudes wire, cuts it, and passes it on. The next one bangs the front end into a flat blade, and bangs the back end to make little anchors. Step three dips the metal rod into a plastic mold to create the handle. Step four applies paint and glazing, and step five polishes and preps the newly minted screwdriver for packaging.

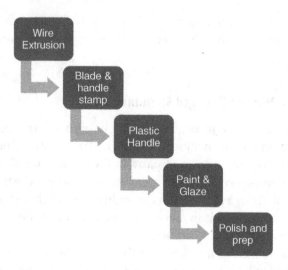

Remember that the classic way to look at this process is that we wanted to run each machine as much as possible, to our magical ROI. What if each machine, each part of the process is a different speed, producing a different number of outputs per hour? What if one of the early processes wasn't producing high quality parts, that then went through the rest of the process?

Obviously this happens all the time, and most pre-Lean factories were a mess of waste and quality issues as a result.

Back to our screwdrivers example, what happens when you run all the machines at top speed, each producing as much as possible, is you get lots of metal rods sitting around, waiting for the plastic molding step, or you get lots of half-finished screwdrivers waiting for the polish step. That gums up the factory floor, costs money in inventory, and pulls focus from what the factory is really supposed to be doing, which is making things people want to pay for.

It has been reported in more than one study that over 90% of construction supervisors think their jobsites are inefficient – clearly some of the same issues that pre-Lean manufacturing faced are going on at our jobsites.

The pre-Lean mindset was to focus on getting the most out of the machines, when it should have been making the most defect-free products as possible. The machines cost what they cost, whether you run them 100% or 50%. That's fixed. But you can control how much inventory you've created, and if you're focused on the whole process, you're going to ask if each machine really needs to be set at 100% speed, or maybe if you can balance things so that each step only makes enough for the next step.

Managers who understand the lean mindset understand that the factory's job is not to make the most of its machines. The factory's job is to make the most of its process, of its total collection of workers, materials, and machines.

The real change, though, is in the role of top-down versus bottom-up planning. You see, once that mindset was changed, it was realized that front line workers understand the day-to-day process much better than managers, so they were given authority to stop the line and make adjustments, set short-term plans and make improvements.

These same workers were collected into small groups, called "quality circles," who would meet frequently and talk about how their parts of the process could be improved – usually this part included several adjacent steps. And finally, tons of data was collected on how the process overall was going so opportunities that improve at that level could be seen and decisions made.

Over the course of the second half of the twentieth century, Lean thinking took Japan from a burned out, poverty-stricken country to one of the leading economies in the world.

Keep in mind, the Lean manufacturing folks never stopped worrying about ROI for their machines. And you'd better believe that no one worried one bit less about safety or the always-present possibility that something unexpected will go wrong. What they did do was expand the tools they knew how to use well, and create some new ones, especially around data.

The Lean Construction example helps us in two ways:

First, it is a great example of how changing mindset changed what managers and workers thought of as most valuable. And when that happened, new processes, new opportunities for improvement came about. In fact, Lean swept around the world and changed what *everyone* thinks is valuable in manufacturing.

Imagine what is happening in construction now – we are slowly adopting a digital mindset, and it is unlocking new ways of looking at everything. Specific solutions and technologies will come and go; what matters is how you, the construction professional, are able to view these technologies and make them work for you. In the case of Lean Manufacturing, over seven decades of technology have changed everything about what happens in a factory, but the core ideas and methods are the same as they were in the 1950s.

How you think about the world affects what you are *able* to do; what you value drives what you *choose* to do.

Secondly, Lean illustrates an important concept – some of the most important tools you use are in your own mind. In the case of Lean, those tools included using data and statistics, but also the habit of asking folks on the floor what was going on, instead of being quite so top-down, command and control. Those same tools are used in Lean Construction, as we'll discuss later.

Mindset is a powerful thing. We are creating a digital mindset that will usher in a new era of construction.

## Construction's New Era

Sometimes we can lose sight of the amazing power of construction. Yes it's a huge industry, yes it employs a huge number of people. But so is retail, so is real estate. Construction isn't just big, it is foundational, and it matters that we continually improve.

Unlike any other sector of human activity, construction makes possible our most basic needs, and our highest aspirations. Construction tames nature, creating the safety and predictability that makes everything else possible. Construction also thrusts our buildings into the clouds, creating spectacles that inspire us to believe nothing is impossible.

Society needs construction, and in a world remade by the 2020 pandemic, we are going to need new things from construction, and construction is going to need new things from technology.

Construction is going to change, probably faster in the next five years than in the past fifty. Of course, it has been evolving and changing all along, but those changes have been slow and not evenly distributed across the industry. For example, the role of architects has changed from that of "master builder" to designer, leaving general contractors to take on risks that maybe they didn't originally want. The information gap that architects left has changed how RFIs[1] happen, how change orders happen, and so on.

Similarly, the traditional design-bid-build, while alive and well for many projects, is under pressure around the world, as Design-Build and Integrated Project Delivery approaches are explored. But most of these changes didn't reach to *how the building is built*. Prior to about 2010, the massive digital transformation that has swept through every industry from banking to farming barely touched construction.

In the last decade, though, things have started to go digital in the field. According to James Benham of JBKnowledge: "It's all driven by consumerization of technology. Workers who have iPhones and apps and can facetime their daughter three states away expect at least that level of technology on the jobsite. When they couldn't find it from tech firms, they started making their own."

That's a part of the mindset shift we're talking about. The change in expectation, the change in sense of digital competence that allowed superintendents and tradesmen who were not necessarily technologists by training to feel like they were good enough to create their own solutions. They felt that way because they use digital technology all day long outside of construction, so it seemed obvious that they should use it on the job. But that consumerization mindset is just that, from the realm of the consumer.

To really make the difference we have seen in other industries, digital transformation requires that the mindset shift from that of construction professionals dabbling on their iPhones, to digital construction professionals working in software and other technologies from the ground up. We need to cultivate the skill of understanding what kind of problem we are trying to solve, and using the right tool, whether intuitive or digital or a combination, to solve the problem and put the work in place.

---

[1]RFIs, or requests for information, are a means of clarifying what an architectural team intended when the plans provided aren't sufficiently clear. They often are necessary when the plans do not contain certain information or specifications.

## The Change Is Coming

Technology is here to stay, and it will become more and more a part of the construction process. As we discussed earlier, specific products may or may not make it, but digital transformation is here and will reach into every aspect of the building process.

The driver of change is what's going on in the world outside of construction. This book was written right during the biggest pandemic in 100 years, and while we won't talk about Covid-19 very much, this crisis has added fuel to existing arguments that the status quo needs a rethink.

Let's look at four external drivers of change: demographics, climate change, owner pressure, and the 2020 pandemic aftermath.

### Change Driver #1: Demographics

In the coming years, Baby Boomers are going to continue to leave the industry because they retire, can't physically do the work, or for other reasons. Boomers are the largest generation in history, except for their kids, the millennials – it is a lot of people. Boomers are typically considered anyone born between 1946 and 1965, Gen X is 1966 to 1985, and Millennials are between 1986 and 2005.

Gen X as a group is a few million smaller than either Millennials or Boomers, which means that Millennials are going to be in decision-making roles much earlier in their careers than would naturally be the case. And the thing about Millennials is that they are not just "tech-savvy" – having grown up with the internet and smartphones, Millennials are intolerant of tech-averse workplaces. For companies looking to grow, or just replace leaving Boomers, remaining pre-digital will have a serious cost in terms of ability to attract and keep young talent.

Demographics are like a slow moving wave that people usually ignore, though they can easily see it coming. And just like a big ocean wave, demographics are inevitable. This demographic wave of technology-demanding Millennials is going to have huge impacts on the construction industry, and not just on the technology adoption front. As an example, Millennials buy homes at a much lower rate than Gen X or Boomers, and we're already seeing that fact change the shape of suburbs and cities, a change that will only accelerate as Boomers retire.

The key takeaway is that, whether it's because they demand it, innovate it, or are in a position to make the decision to purchase it, more and more technology will be used on the jobsite and across the construction value chain because of Millennials.

But demographics have another impact across the world – a radical change in the parts of the world that have enough working age people, and those that have too few. In the past 200 years, every national population has gone through a period of boom, then slow decline as the establishment of basic healthcare and sanitation causes an imbalance between dropping death rates and still-high birth rates, followed by an almost universal dropping of birth rates over the decades to levels that are too low for the native population to replace itself. This pattern is leading to shrinking populations in most of Europe, Japan, and soon, China. Africa, in contrast, is exploding in population, and will continue to do so for some years.

This aging of the population in the north and surging growth in the south will have big implications for where construction happens, how it happens, and how many young people are available in the local market to work in construction. As an industry that faced years of labor shortages, this will be another driver of technology adoption as we seek to augment human workers with technology of varying kinds.

Far from automation taking jobs, in Europe, North America, and much of Asia, automation will save jobs by making projects possible that wouldn't have been possible without it.


## Change Driver #2: Climate Change

In the United States, 40% of greenhouse gases come from the built environment. Globally the number is similar, at 39%.

The construction industry is projected to create 2.2 billion tons of waste by the year 2025. That is a massive number, and it cannot be sustained forever. In recent years, for example, many developing nations have stopped accepting US and European solid waste, forcing hard choices about where to put it. The great Pacific garbage patch, a floating collection of small and large plastic roughly 1.6 million square kilometers in size, is evidence that concern about where we put our trash has gone from an environmentalists' abstract warning to a concrete, practical reality that threatens our food supply, as well as our actual health.

Whether it's from governments, concerned groups, or investors, the pressure to reduce solid waste from construction is only going to grow in the coming years.

Climate change is not just a negative driver, though. While construction does contribute to climate change, it is also our first line of defense against many of the dangers climate change will pose. As low lying areas in Italy, the Netherlands, and in the USA, including most of southern Florida and the gulf states of Alabama, Mississippi, and especially Louisiana face the prospect of rising seas, it will be up to construction to build the walls, levees, and drainage that will save our communities from the storm surges that increasingly threaten them. It is easy to look at projections of whole areas of the country underwater in 50–100 years and be dismissive of these projections as uncertain. But we cannot overlook the fact that it isn't required that the land to be underwater for there to be a threat – hurricanes like Katrina and Sandy showed how much damage a two-day storm can wreak with oceans right where they are now.

## Change Driver #3: Demands of Owners

Owners and developers have become more and more demanding in recent years, whether it be for increased safety, or more recently, workforce development. Organizations like the Construction Users Round Table (CURT) often lead the way in driving new standards, and certification mechanisms that lead owners to require higher levels of effort to achieve these certifications. And as more and more technology companies become owners of built environmental assets, they are demanding the use of project and field management technologies that are on par with other areas of their businesses.

## Change Driver #4: 2020 Pandemic Response

Within weeks of the lockdowns caused by the Covid-19 pandemic, construction companies and their technology providers were adjusting to the need for social distancing by conducting many more meetings remotely, and were beginning to apply on-site sensing technology to monitor social distancing.

Companies like StructionSite, that provide 360° photo management of job sites, provided their services free or at reduced fees for construction clients who were adjusting to social distancing, and who were suddenly less keen to send non-essential staff anywhere. Similarly, like everyone else, construction teams have become much more willing to hold meetings over digital teleconferencing software. After months of this, many have permanently changed workflows and processes to be more digital and remote.

The length of the social distancing requirements during lockdown and gradual re-opening has meant that many processes will change forever, and the coming years will see the effects of that change as new features on existing products, as well as entirely new products, are introduced to address these changes, and in some cases make new workflows possible. As an example, we expect that virtual and augmented reality will both find swifter acceptance and adoption in a world where being together physically is much less valued than it once was.

Across industries, we have seen company after company report that the Covid-19 pandemic has led to a decade of technology adoption happening in a few months. It has caused cherished assumptions, like the need to be physically in the same space, to be challenged and proven wrong, making other assumptions open to re-evaluation. We cannot yet know the final impact of so large a change, but Covid-19 has already led to big changes in construction's use of technology.

## Technology Supply

At the same time that these factors pressure construction industry players to adopt new technologies, recent years have seen a flood of new technology products, especially software. Everything from new ways to create, share, and use Building Information Models (BIM), to a flurry of project management solutions, to drones, voice-operated software, and much more has suddenly become available. Along the way, the venture capital community and other investors have "discovered" construction technology, leading to more and more options.

While this flood has led to some companies feeling like it's too much too soon, construction companies have created teams focused on successfully navigating all the options, adopting these new technologies and making them work for their teams in the field and back in the office.

And with all of these folks pushing to create new solutions, some genuinely useful things are coming to market. It is no exaggeration to say that products like Procore have transformed how work gets managed and handled on the jobsite. Workflow and jobsite capture tools like Holobuilder change the level of awareness companies have of their jobs across space and time in a truly meaningful way.

The past few years have seen an evolution from clunky, disconnected point solutions to more networked approaches where many software solutions do in fact work together. However, truly connecting every software product together remains a difficult problem in construction, with many companies either hacking together solutions, or suffering through the painful double and triple data entry issues that have plagued the industry for years.

All of these solutions are coming at the construction industry because they, or something like them, are already being used in other industries – construction has, for valid reasons, been a slower adopter of technology for most of its recent history. The benefit of being second – or let's face it, more like a distant fifth – to the technology party behind other industries is that construction firms and workers are able to learn from decades of thinking about digital transformation and cherry pick the best ideas.

This book will help you do that.

## Use of technology *is* a technology

Every time you use a powertool, or a non-powertool for that matter, you are applying knowledge, and that is a technology. Processes, know-how, these are all a kind of technology, they are a means of getting something done for a purpose.

Every new tool requires that you learn some things. A new kind of hand tool takes some practice, and real-world fiddling around to understand how to use it. This is, again, a kind of technology. For these sorts of non-digital tools, though, there's really no point in calling it "technology," it's just called "know-how."

The reason I call out this definition of technology, as the *how* to do things, not just the *what* to use, is because unlike non-digital tools, software and digital solutions work in very predictable, narrow ways that you must learn in a very different way from how you learn a non-digital tool.

No one would suggest that you learn to use a drill without ever holding one. But you can learn a *ton* about Procore without ever getting on the platform, through their eLearning. You can learn everything you need to know about Revit from tutorials and videos, and the same is equally true for almost all software-based products.

The technology for learning is, I think, one of the biggest changes coming to construction. This learning is as easy as watching a YouTube video, and can be done from a phone, desktop, or tablet anywhere.

In the past, workforces have been encouraged to "adopt" things as if it were just one big change, then you'd get back to a steady normal. And maybe that was true in the 1970s, but it just isn't true anymore.

Software doesn't get updated every few years and get shipped to you like Windows in the 1990s. It is a website hosted somewhere else, in the "cloud." It is constantly updated, redesigned, replaced. The Prolog of the 2010s is the Viewpoint of today, and the Procore of tomorrow isn't going to look like the Procore of today, because it will keep evolving.

This poses a risk for the everyday Joe and Jane who don't really want to focus on technology, but actually love doing their real job, whether that be framing walls, setting up mechanical systems, or stringing up electrical systems. The solution is to understand the basics of how software, AI, BIM, Prefab, and other technologies work.

## The Rest of this Book

The next chapter is devoted to software, from the ground up. You might know some of the concepts, but probably not the frameworks introduced. How software gets made will make you a better consumer of software, as well as a better selector and, it is my hope, a co-creator of software products.

Chapter 3 discusses the more modern aspects of software: networking and systems of software. This includes the cloud, APIs, and other terms that are essential to understanding how the software you use, from email to Procore to Revit, connect to each other and operate.

Chapter 4 goes out into the field, with a deep discussion of construction management software and a wide range of other field software products.

Chapter 5 looks at industrialized construction, which is made up of automation, prefab, and manufacturing. Automation specifically is a boogeyman for many field workers, as there is a concern that robots either in the field, or at an offsite manufacturing facility will erode wages or take jobs altogether. We'll explore this from a technology as well as an economics standpoint.

Chapter 6 looks at the ultimate technology boogeyman, artificial intelligence. As the most hyped technology in many years, AI has suffered from too many stories about what it *might* do, and not enough practical discussion about what it currently can do. There are hard limits to what current approaches to AI can handle, and a huge gulf between those approaches and any of the nightmare scenarios that have so obsessed some in the media.

We extend those ideas into Chapter 7, with a review of applications of AI, things you can actually pick up and use right now and in the near future.

In Chapter 8 we'll get into some of the sexier technologies that are making their way into construction, including virtual reality, augmented reality, the Internet of Things (IoT), and more. These technologies promise to completely change how projects are imagined, planned, and built.

Chapter 9 explores technology innovation and its adoption by companies and users alike. Well over 100 years of experience and research into how technologies spread and more recent work on how companies can most effectively manage technology adoption as part of a digital transformation strategy give us a strong foundation for charting a profitable path forward.

Finally, Chapter 10 will summarize the book's core themes of mindset changes and attempt to predict what construction might look like in 2035. Such a prediction will be wrong in the details, as they always are. But we know enough to make some bets that can help you decide on what to do, and learn next.

# CHAPTER 2

# Software

> "Software is eating the world."
>
> – Marc Andreeson, "Why software is eating the world," *Wall Street Journal*, August 20, 2011

A cross the economy, industries have digitized. Firms have turned analog, paper-based processes into software-centered approaches, to the point where innovation in software is more important to the success in these industries than anything else. From retail to manufacturing to travel, software has become the arena for competition and efficiency improvements.

This trend achieved peak visibility with the digitization of transportation with Uber and Lyft, and hotel rooms with AirBnB, along with a huge number of other "gig economy" jobs.

The same digitization trend has been happening in construction for years, though it has been a bumpy ride. Many in the industry will tell you that software first entered the construction industry through accounting software, and was often not very easy to use. Recent years have seen the rise of more modern solutions like Procore, BIM360, and a ton of more focused solutions like Rhumbix, Raken, and eSub. What we know for sure is that software does add value; it can even transform industries, even if it takes some adjustment.

At the same time, lots of software is a pain to use, difficult to integrate, and doesn't always do what we need it to. Finding out that promised benefits don't really materialize, or that the software flat out doesn't work, takes effort and resources. Understanding how to assess and use software is critical.

When we talk about "technology in construction," the first thing most people will think about is software. As the quote above suggests, software is now everywhere, from the tools in your hands to the schedules on your laptop to the 3D designs changing how we conceive and build our world.

This chapter is going to introduce software from the ground up, giving you the tools to understand what it can and can't be expected to do, and how to ask the right questions to make sure you are able to master the new toolkit that is evolving. We begin, as we did with technology, with basic definitions, then go on to cover key concepts that make modern software work and that you'll need to know to understand how modern software does its job.

## First Principles: What Is Software

Just like some of the other words we've dealt with earlier, software gets thrown around a lot, without anyone taking the time to really define it. Let's do that here – I like the Wikipedia definition:

> Computer software, or simply software, is a collection of data or computer instructions that tell the computer how to work. This is in contrast to physical hardware, from which the system is built and actually performs the work. (Wikipedia)

Three things jump out from that definition: data, instructions, and hardware. Think of data as material, instructions as process, and hardware as ... hardware, and you have a machine just like anything else. Data is what software works with – instructions are written in code that define what the software can do, just like gears in a physical machine – and hardware is the machine, like an iPhone or laptop.

Whether software is running on your iPhone, laptop, or on a server somewhere, its function is to receive information from the world, translate that information from the way it occurs in the world into the way computers need to deal with it, pass that computer-friendly information to the hardware, where the instructions that were programmed by some group of people are run, then pass back out the results to the world, usually showing up on a screen somewhere.

The business of creating software has a lot in common with creating a building. There is an owner who decides they need software built, and they hire an architect to design the software, specifying some but not all of the details. Engineers then come in and design most of the rest of the software, with specialist trades filling specific things like user experience, web presence, and API connectivity.

Software has always been seductively incompetent. It does a few things extremely well, but not other things we usually think it should. Every person I've ever asked has a story of software failing them somehow. Modern software, because it is usually changed almost constantly, is often more competent than software of the past, but it is still by definition very limited.

Earlier we covered how mindsets matter, and how we think about something affects what we are able to do with it. Thinking of software as a machine, built to do certain things, but only those things, helps us to understand how to avoid the frustration of thinking software is more than that.

## Software Is a Machine

Software is like any other machine you use to do your job. It is a set of tools that allow you to do more than you can do without it. It should make your job faster, allow you to automate some things, or at least make you safer.

In the coming years, software is going to continue to get smarter and smarter – a point we'll cover even more in the artificial intelligence chapter. For now, though, let's talk about the limits and possibilities of software.

As we discovered earlier, software is a set of instructions. Someone had to think of what it should do, write those instructions, and publish them. These instructions are exactly like gears in a machine – they allow the software to do a sequence of things in response to some action of the user. But just like gears in a machine, software cannot do one single thing that it isn't designed to do.

This can trip us up, because almost nothing else in the world is quite that strict in its limitations. Physical machines have humans as part of their operation who provide a critical flexibility that software doesn't include, and of course actual humans are inherently flexible and able to adapt automatically to fuzzy inputs, unclear situations, and new scenarios. People might only be qualified to do the jobs they're trained in, but without much fuss they can help out in other areas, even areas they don't understand very well. Software doesn't do any of this – unless there is a human in the loop somewhere, software cannot adjust to fuzzy commands or new situations at all.

This matters because when we create software the first thing we always do is sketch out very specific requirements. And you as the buyer or user of software need to do the same thing.

## How We Make Software

In fact, the way software gets built reflects this need to get the requirements right. The first step of course is to decide something is needed, so you sit down with the folks who'll use that something and decide roughly what they need and want, and what functions will be required to do that.

That'll be where you decide what kind of software to build, whether it's a mobile app or cloud-based app, whether it needs a big database, and so on. At this point, a good process is to create a prototype. Prototypes are important because they give you something tangible to play with, and the mind always does a better job of understanding things when there is something tangible. Keep in mind, a prototype doesn't do all the functions you'll eventually build, but it looks and acts like the software will. Check out www.proto.io, www.balsamiq.com, or www.invisionapp.com if you'd like to see some examples of great prototyping tools.

Next, the prototype must get tested with potential users, before anything gets built. This is critical and too many startups skip this step because they're in a rush or don't know how to find testers. But if you don't test with users first, you cannot be sure the product is needed, that real people will understand what it's for, or that you've got the design direction correct.

After we've tried out the prototypes both internally and with users, we go back to the drawing board and make changes based on what we've learned. This usually has to do with what buttons to click, what sequence of interactions are more natural, and what features might be good to add.

## Agile Development

What happens next is what we call the Agile methodology. This is actually how most modern software gets built, and it is important for you as the construction technologist to understand how it works and why it works that way.

In the past, software would be planned like a building. Every feature and function would be mapped out, scheduled to the day, and packaged up in what they called a "waterfall plan," because each step in the process cascaded to the next in an orderly fashion.

In practice, this meant that the team doing the planning would work up their plans, then send the engineers away for months at a time. The engineers would come back with their shiny new software, usually later than intended, but most importantly, that software wouldn't come back the way the planning team thought it would.

It turns out that, just like in construction, there are countless little decisions that get made along the way when you're building software, and if these are just made by the internal team, eventually, bit by bit, you get away from the first intention. Unlike construction, there is no contract that specifies a plan that is the criteria for delivery, there are no RFIs as these decisions get made – the software team just makes the best decision it can and moves on.

This approach caused delays and bloated expense, and most importantly, ensured that key decisions about the final product weren't made with either the project owner, or end users in mind.

So, in February 2001, a group of top engineers published a new way of thinking about developing software, in the Agile Manifesto. At agilemanifesto.org you can see the 12 principles of that manifesto. It focuses minds on creating working software early and often, and getting business feedback all along the way.

In the years since that publication, lots of thinking and processes have grown up around the agile process, and it's settled into a few flavors. I tend to run a lighter version of Agile because the teams I've run aren't quite at the Facebook or Microsoft sizes.

The process begins either during the prototyping phase, or right when you've got your user feedback together. There's a kickoff meeting, where the schedule and roles are defined. This helps and isn't so different from normal projects.

Then we agree on a cadence of client meetings, usually every two weeks. These two-week periods are called sprints, and in the beginning it is agreed with the software developers that there will be a set of requirements they will work towards, and that this spec won't change during this two-week sprint. This allows the team to focus, and speeds things up considerably.

These sprints must produce working software, and the client needs to use the software and provide active, concrete feedback. This does a few things: first, it stops developers from getting lost in huge blocks of code. Second, it ensures that the client is able to adapt their requirements to real software, not the abstract idea or prototype they started with. And finally, it keeps the team working together towards a mutually understood goal.

Between these sprint meetings are daily meetings of the team, called "standups." The point of a standup is to avoid the long, drawn-out meetings that so commonly happen – you actually keep everyone standing so it naturally lends itself to being a shorter meeting. Each person talks about what they're working on, any issues that have come up, questions or needs. Then everyone goes to work.

The point of Agile, then, is to keep a tight connection between what the client or user needs, and what the team builds.

## Software Anatomy

We think of software in layers, so here we're going to go through how it works, and what the key pieces are, from the screens you touch and interact with, down to the hardware that drives the whole thing. This section is purposely arranged to be easy to refer to later, and will be referenced in the index for that purpose.

## Users

Software development refers to anyone actually using the software as "users." This could be a superintendent in the field looking at schedules on an iPad, a sub-contractor doing a daily report on their phone, an estimator working with a spreadsheet in the office, an architect designing the overall building, a specifier working in Revit to detail a bathroom, and many others.

What users will do is notoriously difficult to pin down, because most of us don't think the way software operates. We're used to dealing with other human beings, with minds that can figure out what we mean when we're not that specific, clear, or complete in what we say.

In contrast, software is really bad at guessing what you mean. Remember, it is just a collection of instructions, despite how professionally designed and slick it might look – at the end of the day it is just a "machine."

As an individual user, sometimes it can be hard to see quite how used to other humans' common sense we are. People, even people we might think aren't paying attention, are amazing at interpreting what other people mean, and can make up for tons of ambiguity by understanding what's going on, what the context is.

Every user is in a context. They're in a specific city, on a specific job, doing a specific task, with a specific set of needs and facts to deal with. Any human will either already know, or can guess, most of these contextual points. So we as humans just take for granted that software will be able to do this, because every other interaction in our lives is with a person.

Software has no idea about context. It only knows what it is designed to accept from you, which is usually a few button clicks and a few typed words.

A good analogy is customer service at a hardware store. The customer service rep knows nothing about you before you get there, all they have to work with is what you give them. Sometimes they get your problem quickly, solve it, and off you go. But what if you're missing a receipt and they have a receipts-only return policy? That customer service rep is only able to hear one thing: "I want to return this, here is the receipt." No amount of stories will change that, the customer service rep has zero interest or ability to do anything with your context. You either have what they need to do their job, or you do not.

Think of that analogy the next time software you are using gets stuck – it isn't your fault that the software can only deal with a narrow range of inputs, but it is still the reality. The art of user experience design is anticipating needs and creating software that can accept almost all of the ways most users will try to interact with the software, but they'll always miss something. We call those "edge cases," the times when a user does something that the software can't handle, that most people don't do. The hard part about edge cases is that the more complex the software, the more of these unanticipated, unorthodox ways of interacting with the software there will be. As a result, the more complex the software has to become to deal with these edge cases, and so on.

About two years ago, I was developing a product that used voice as an input for construction jobsite reports. I'd created a prototype that had the computer call users and ask them a series of scripted questions. To make it faster and easier, many of these questions required simply "yes" or "no" answers. The first question the computer would ask was "Was the job delayed today?" I'd programmed the system to accept a "Yes," or a "No." Seems obvious, right?

I had a good friend of mine, Damon Hernandez, test the system early in the development.

Now Damon has been building software for many years and wanted to help me improve the software. And Damon has a sense of humor.

So when the system called, Damon answered as follows:

"Hello, would you like to do your daily reports today?" asked the
    eager AI.
"Yes," said the patient Damon.
"Great. Were there any delays on the job today?" asked the still
    chipper AI.
"A little," said Damon.
*Silence ...*

Just like that, Damon showed how many ways a real person will interact with software, because we're used to speaking to other people, and other people are much more flexible and able to handle edge cases than software is.

*All software misses some edge cases.* So if you're trying to get something done with a new bit of software, often it means you're just giving it inputs it's not built for. So you can either try it a different way, or Google how to do whatever you're looking to do. I usually suggest the latter – in fact a theme in this book is that whatever trouble you're having with something technical, someone else has had that problem too.

*Users are not "alone."* There are groups all over the place who are trying to learn, use, and optimize software. You can Google them, or you can go to the software or hardware company's website and look for the Frequently Asked Questions (FAQ) section, and the user forum they often host.

As a software developer myself, I can tell you that creators of software learn early that no question is stupid; in fact, the stupid questions get asked the most. Googling or going to a forum is perhaps a new way to ask advice, but it is really no different from asking a friend – in this case anonymously if you want, and usually pretty fast.

## User Interfaces (UI)

Any way a user can give information or input into a computer, and any way it can get information or output, is called an "interface." You will sometimes hear this referred to as "UI," for "user interface."

There has been an incredible range of interface types. Back in the 1960s and 1970s, the interface was, believe it or not, punch cards. Users would put punch cards into the computer, it would read them, do whatever was needed based on the information in those cards and the computer's own software instructions, and then the computer punched its outputs on either the same card or another one.

As crazy as that sounds, it helps point something out – user interfaces are driven by the technology under them. They will only be as good as the software they connect you to, so great software and great hardware usually have great interfaces, and limited software/hardware usually have limited interfaces.

In the case of punch cards, computers back then had effectively no memory, and could only do one operation at a time, so the punch card had both its data and its instructions in the same place. There's a reason everyday folks didn't hear very much about computers until the 1980s.

That's when computers got powerful enough to use screens, and had their own memories, small as they were. As the 1990s progressed, our computers got better and better, and we went from typing commands to the graphical user interface (GUI) that we take for granted now. But when Apple introduced the Macintosh with its image-based screen, followed by Microsoft introducing their Windows operating system, it changed the world. This was when we started being able to use pictures to represent things, like the trash or a given file, and we can drag and drop, among other things. It made computers useful for normal people.

Back in the 1990s, most people only used a few applications, like Microsoft Word, Excel, and some PowerPoint, so the need for really intuitive user interfaces wasn't as high as it became later.

Then in 1993, a recent University of Illinois graduate named Marc Andreeson took the idea of "hypertext markup language" that had been invented by a British computer scientist named Tim Berners-Lee, and created the Mosaic browser.

As the Mosaic web browser gave way to Netscape, and AOL and CompuServe took many, many people online, the world wide web changed forever how we deal with software and applications. Suddenly, we had all these new things to do, like writing and sending email and browsing webpages. In time, more and more offline tasks went online, from shopping to planning trips to paying taxes and so on.

This drove a huge increase in the importance of user interfaces, and because there were so many new applications, we had to start thinking about how users clicked, typed, and watched video in software. What were they looking at the most? What did they ignore? What key instructions did they miss?

Earlier we explored how people are used to interacting with other people, so they don't naturally ask what rules are necessary to make themselves understood. Over the course of a lifetime, every time someone had to get something done, at the other side of the interaction was a person who was amazingly good at interpreting what they meant. And software just can't do that.

The very first websites were pretty simple, but as they became more complex, designers kept running into this issue where users would do things they hadn't anticipated, and we got better and better at designing for real people.

## User Interface Design

Along the way, the millions, then billions of people clicking on products and websites started to generate huge amounts of data. In fact, because this data was so random and hard to predict, it required a whole new class of tools. These tools became "big data," which sim- ply means data that is not classified into neat rows and columns, but is instead just tagged for analysis later. Big data is how we deal with an unpredictable stream of clicks and swipes and button presses and video viewing times and so on. We just let users do what they want, record it, and store it later.

Big data allowed user interface designers to collect unexpected data and see patterns they weren't specifically looking for. In contrast, older data formats assume you know what you'll be getting – so they have neat rows and columns, like a bank account or your grades in school.

This huge quantity of data required new ways of looking at data, and two strategies are used to deal with all of this new complexity and user behavior data: design standards and testing.

Design standards are really just lessons learned, often combined with some visual style guidelines. We've learned, for example, that most users in the US start looking at any website or product interface in the middle of the page, and then to the upper left. This has been ingrained from reading paper, and still works.

We know that people are attracted to movement on the page, and that messy pages make people feel stressed. There are a bunch of other things that experienced UI designers know, and they create the starting point for any product's or website's design.

To work, a software interface has to have created data that it needs to send into the software itself, so the software can understand what you're telling it to do with that typing, clicking, or whatever way you're interacting. That data might be permanently stored, or it might just be used, then disappear.

When software designers want to understand how well their design is doing its job, they collect this data, often in huge quantities. In extreme examples, big tech companies like Amazon and Yahoo are known to run thousands of UI experiments a day. They move a button, change a description, change some colors, and so on. All of this to make the experience better, and ultimately to make sure that more users buy or click on advertising.

As a user of software, it is important for you to understand that most good user interfaces have been tested a great deal. For enter- prise software, especially new software from a young company, the testing might be much more limited and more like a series of interviews, and what we call "alpha" testing, where a small number of users gets early versions of the software to try it out and give feedback. Later, there will be "beta" testing, which is about fine tuning both the interface and the reliability of the software.

Companies will keep tracking forever, hoping to keep fine tuning interfaces and introducing new features and ways to interact. You can often opt out of this tracking, usually in the "preferences" section in the "settings" menu. At the same time, you can also very often reach out to software makers who you rely on and tell them what you'd like to see, and what's not working. Not every company is set up to handle this, but the good ones are constantly looking for feedback and ways to fit their users' needs better. I'll tell you any software startup should be eager for feedback of any kind, especially from users they don't already know.

## Types of User Interface

We've come a long way from the punch cards. Today, you're able to interact with software through typing, clicking, dragging, touching, pinching, moving your phone, speaking, and more. These interface options should make things clearer and easier to use. Often they do.

However, the experience of some of these user interfaces isn't always quite what we're used to in older, more developed interfaces. For example, most graphical user interfaces are pretty functional and get most of what you want done easily, while newer ideas like voice interfaces still have a way to go.

Voice has been around for years, but only recently has it been anything other than a pretty finicky note-taker. Really useful voice interfaces started with products like Apple's Siri, and were released a little before the technology to do a good job was quite ready. Those of you who started using Siri when it was introduced in 2011 will agree it was a rough start. In the years since, though, advances in artificial intelligence and a mountain of user testing has made voice user interfaces much more intuitive and powerful.

Similarly, anyone who's tried augmented reality interfaces on the Hololens or similar devices will agree that getting things done in AR still takes some getting used to. The underlying technology is quite complex, but we're just not used to "clicking" in mid-air, and it shows.

I personally think that augmented reality is going to use voice much more than other products do, but we'll see.

Layout and operation of the user interface is the first thing you learn with new software, and it is most likely the source of whatever early confusion you might have. That interface was designed to make your life easier, so if at first you don't get it, jump onto YouTube and look it up. My experience is that once people figure out what the software designers had in mind, it's a quick process to figure out the rest.

## User Experience (UX)

The user interface is the most obvious part of the overall experience of using the software to get the job done. We call that broader experience "user experience," or UX. This is the total experience you have as a user of the software, and has to do with things like sequence of actions, choice of which functions to put where, and so on. The difference between user interface (UI) and user experience (UX) gets confusing, but you can think of it as what you see (UI) and what you do (UX) with software.

The experience of using software starts with the first time you decide to try it out, through to your learning how to use it, to the actual use of the software, especially when you're using that software with other software.

Different companies choose to look at UX in different ways, with some looking to make the experience very positive, often searching for "delight" amongst users; these are usually consumer software companies like Apple or Facebook.

Other companies are much more hardnosed and looking to make the experience efficient, easy to understand, and useful. Not surprisingly, this is the approach many enterprise software companies choose to take.

Users of consumer software are also users of enterprise software, so the need to have well-designed UX with a little personality can bleed into enterprise software as well. The senior superintendent logging into his iPad to see the latest drawings is also logging into Facebook to see his daughter's field trip. Very often, enterprise UX is judged against consumer UX, and therefore has an influence.

Just like with the user interface, good software companies want to hear how your experience was, whether you were able to get from one area to another, whether you understood which steps to do when, and how well the software recovers from errors either you make, or someone else makes, because this happens all the time.

As a user, and sometimes buyer, of software, I would not recommend you rely on UI, or the design of the software, to make judgments about whether the software is right for you. What matters much more is the user experience, especially how well thought out the menus and graphics are, how well the system connects across different components, and how intelligently the steps to use it are thought through.

When you're looking at software, the company will do a demo, and it will work easily, of course. When you try it yourself, you should expect to be a little confused at first, but then see how long it takes to figure it out, and see how helpful the software's onboarding videos or web-based tutorials are.

Always remember for this new software to work at any kind of scale, lots of people will need to learn it pretty quickly. So the learning curve should be fast, and the software company should have plenty of materials to help.

Despite the obvious need to make onboarding seamless and easy, many companies are pretty bad about helping new users learn to use their software, beyond some very high-level functions like logging in. For more established software, other users are going to be your best source for how to get the software to do what you want, and to some extent, to understand how good it is, though I think most people would agree that the internet is hardly an unbiased source of opinion.

An exception, and a good example of well thought out onboarding is Procore's "certification" program, where you go through about 90 minutes of a video tutorial that is not meant to be difficult, but is exhaustive in going over every function you are going to need. Procore incentivizes users to do this tutorial by providing a badge that can be displayed on LinkedIn. So they've managed, with this tactic, to both provide in-depth user training and also get people to brag about supporting Procore.

## Piloting Software

So far everything we've covered is about how easy software is to use, but in the end you are buying software so that it will have a beneficial impact on your business that is big enough to be worth the trouble and expense of a real deployment. The only way to know that is with a pilot.

Many companies have an innovation department that can fund formal piloting programs, which is essential. The innovation team should develop criteria for assessing new software, from onboarding to user support through to actual effectiveness on the job. Wherever possible, it is better to run a pilot, even for an established software product, because every software product will change your processes – understanding that impact is critical. These impacts are often driven by the user experience, which in turn allows, or prevents, crews from using the product successfully.

An important area where user experience continues to advance is how intuitively the software responds to, and even anticipates, what the user will need next. Established software, like Autodesk and Procore, is usually better at this, as they have had time to build out all of the menus, options, and extra functions that ensure almost no user is stuck, unable to do something critical to getting the job done.

We're starting to see more and more natural interactions via voice that leverage artificial intelligence. These user experiences are just beginning to reach what I believe is a transformative level of ease and efficiency for users.

However, because voice lends itself to users assuming they're really talking to a human-like counterpart, we are again seeing user frustration as software isn't always able to understand what users are saying, especially when they use uncommon idioms or ask unexpected questions that a human could handle easily. Underlying all of these products are data-driven AI systems that learn over time, so just like Siri today is dramatically better than Siri of 2012, voice interfaces for construction software will soon make a huge difference in your ability to use them efficiently, especially in the field.

## Applications

User interfaces are how you tell software what you want it to do, the user experience is how that software works for you in getting your job done. The software itself is called an application, or app.

Until the mid-2000s, most software applications were delivered to a user via some kind of storage medium, like a floppy disk or CD-ROM. It is easy to forget that in 2004 there were tens of millions of users in the US that connected to the internet via a dial-up modem, so the internet was not a good way to distribute software. For comparison, a dial-up modem had a top speed of 56 kB per second, versus modern Wi-Fi speeds upwards of 2 MB/s. Many of the applications you use started their lives in an internet environment 35 times slower than the slowest thing out there today.

As broadband became normal, a new kind of software application started to take hold. Instead of sending software to companies that they then installed on their own computers, or "on premises," software started to be hosted on remote servers. Those servers are usually housed in data centers, along with thousands of other servers, and they can be "rented," so you don't need to have all that equipment yourself. This collection of servers became what we now know as the "cloud." The cloud enables companies to put applications on servers that sit somewhere else, that can then be accessed remotely; thus those applications can be sold not as one big tangible item, but as a subscription.

That subscription is what we call "Software as a Service," or "SaaS," and it has transformed the technology landscape in some very important ways.

## SaaS Revolution

Software used to be quite difficult to sell in an à la carte way – you bought it all, and maybe you paid for "seats" but most of the time you just bought it, created one user, and then got to use everything the software could do.

This made it harder for smaller companies to use enterprise-grade software, because many of the costs were up-front. An example of this would be Adobe Photoshop, a very common design software for graphics. In 2010, for example, Photoshop was $700 as a stan-dalone software product. Now, you can access Photoshop for under $35 per month for a single user. That is 5% of the price to get started, though you will ultimately pay more if you keep using and paying monthly. Many companies and users that couldn't afford the one-off cost can pay monthly. Since the monthly fees add up to more than the one-off cost, software companies can afford to spend more on product development, which most do.

Beyond cost, though, software that is delivered from the software company's servers means it is still on those servers, and that means the software company can make changes easily. In the past, every change got rolled into a new "release," and those happened pretty infrequently, usually once a year or less.

In today's world, companies release new features constantly, and are able to change the UI and UX on the fly as they learn from users. It also means, though, that it is much easier for a user to switch to a competing software product than it was in the past, though it is still tough to untangle processes and data.

When software is delivered over the internet as a service (SaaS), the software company can give you much, much more control over how much you use, who can use it, and many other cost drivers.

SaaS also means that the same application, and the same docu-ments and data, can be accessed on any computer and very often, tablets and phones.

## Downloaded Applications

As powerful as SaaS is, one issue is always internet connectivity. SaaS usually requires a pretty good connection, and these don't always exist outside of offices and homes. As a result, mobile applications, or "apps" still rely on downloaded versions for most software products. It is not uncommon, for example, for a SaaS product that requires no download for a desktop or laptop to have a downloadable app for the phone or tablet.

This same logic can apply to applications that have heavy com-putational requirements, especially for graphics.

For example, most design software has some downloadable ver-sion, because graphics generally take a lot of computer power, and that's best done locally on the user's computer instead of being sent over the web. You'll also see this for virtual reality, augmented reality, and many artificial intelligence software applications for the same reason.

In fact, one of the reasons there is excitement about 5G wireless networks is the hope that some of this heavy work can also be offloaded and be done in a centralized, cheaper SaaS fashion. As of this writing in 2020, 5G had not yet made it past a few test cities, and the timing of a nationwide rollout with really good coverage was still uncertain.

## Security

Applications are also the front door to the rest of the world, and where security breaches typically happen. The way modern software works is so interconnected that it is sometimes hard to imagine how a security lapse in one place might impact far away databases and networks, but it really can.

Some readers might recall in 2018, when Target reached an $18.5 million settlement for a massive data breach, where 40 million credit card records were stolen. It turned out that their HVAC vendor had had their credentials stolen, which then allowed hackers to get into Target's main computers and steal the data.

Security breaches are difficult to really report on, because so many go unreported – but it is safe to say every company you know, especially large ones, has had a breach of some kind. Especially common are data ransom schemes, where a hacker will lock up the company's data, and require payment to unlock the system.

Because most applications are connected to each other through either local or internet networks, applications you use for seemingly innocent things like sending pictures to your kids can be a danger to your entire company, and sometimes, your clients.

While this isn't a book about security, this has become such a big issue that everyone looking to understand and operate construction technology needs to understand a few key rules:

1. Never click on or download anything from an email without first checking that the user is who they say they are. This can be done by clicking on the username in your email client to make sure there isn't some obviously fraudulent email address.

2. Never send log-ins or any other credential via email, text, or other means. If you have to send these for some reason, split up username and password then send them each in a different way, like text for one and email for the other. Best is to put the credentials in a Word file, use the "Protect Document" function, set a password, and call your counterparty with the password.
3. Never send money without calling first.
4. You can find more tips like this at the FBI's website. Because many of the hackers behind these breaches are foreign, the FBI is putting serious resources into helping smaller companies to secure their systems and their behavior. One place to start is their "recent scams" page: https://www.fbi.gov/scams-and-safety/on-the-internet

## Operating Systems

Applications do not work on their own, they need something that will connect these many applications to the hardware they run on. There aren't very many you are going to need to know about, and they can be grouped into three general types: computer/server, mobile, and open source. Let's quickly review what these things mean, and what they mean to you.

### Computer Operating Systems

Everyone is familiar with Windows and MacOS, these are the two operating systems that 99% of the world's laptops and desktops use. These tell the hardware what to do, and allow applications to use the hardware, like its speakers, memory, and of course, processors. Most of what a computer's operating system actually does is pretty technical. Windows and MacOS do a lot of the same things, but have chosen different ways to do them, and are built with different designs.

Modern operating systems do a lot of things that used to be applications, like connecting to Wi-Fi, running virus checks, and more. If you don't really know your way around the operating system beyond just finding your files, I highly recommend you spend an hour or so with a video tutorial to show you tips and tricks, especially about security. Both MacOS and Windows have a ton of features that you'll find useful – and both have a voice-activated assistant, which is a free way to try this new technology.

## Mobile Operating Systems

Apple's iOS and Google's Android both do many of the same things – they allow your phone to be a computer, operate the camera, connect it to the internet, and more. As the cameras get more sophisticated, and with the addition of a light version of radar, known as LiDAR, to iPads in March 2020, augmented reality will become more an everyday reality; iOS and Android will continue to enable applications that use these features.

Mobile operating systems, especially Android, are also used to run other devices that are mobile and have some of the characteristics of a phone. A good example is the Oculus Quest, which is a virtual reality headset that came out in 2019. Its operating system is a version of Android, which allows it to benefit from many of the applications and innovations that Android introduces.

## Open Source

The third kind of operating system actually overlaps with the other two. While Windows and MacOS are both completely controlled by Microsoft and Apple, respectively, there has been a move towards more open software in the past few decades. For example, while Android is owned by Google, it is a much more flexible standard that allows companies and other groups to make their own versions of the operating system, which is helpful when you want to use it for new hardware, like VR headsets.

Perhaps the most well-known open source operating system is Linux, which is often used for servers and more technical use cases, as it is neither designed nor supported as a consumer-level product.

## The Stack

We started with the most user-facing idea, the application. With operating systems, we have moved closer to the actual hardware, and now is the time to tie it all together with an important concept when thinking about any technology – the "Stack." We call things technology stacks because different elements work at different levels, from the hardware up to the user.

The stack is necessary because users have no idea how to talk to a microchip, and microchips require very specific language to work – what we call machine language or binary. So we need to translate from one layer to the other, but also preserve the ability to have different types of applications.

At the bottom of the stack is the hardware it runs on. That, in turn, usually means a microchip of some kind or another. These are absolute miracles of design and production, where billions of transistors are packed into a space that's smaller than a quarter. Let's cover what a transistor is, and what they do.

## Transistors and logic

A transistor is just a way to use electricity to store a value of a 1, or a 0. It does this by having a wire connected to another wire, and if the first wire fires an electrical signal, it will cause the second wire to either open, or close a gate. What that really means is the electricity from the first wire makes the second wire unable to transmit electricity, or changes it and makes it able to conduct electricity. That's a 1, or a 0.

Now imagine a chip with 1.7 billion of those, and try to imagine how your mac or PC goes from you typing English, to somehow telling those 1.7 billion little gates to do something useful.

That sort of thing happens in all software and communications, where there are layers of different software that translate from what the computer needs to do its job, up to what humans can deal with, and then usually up again to what works best for actual human work and entertainment.

The stack usually starts with the hardware, which is built to work with a specific operating system. That operating system then has applications you can use directly, and these applications sometimes support even higher level applications – for example, a web browser that serves up Procore.

## The Software Developer's Stack

In addition to the user stack we just described, there is another use of the term "stack" that you might come across.

It turns out that applications can often be built several ways, using different sets of software languages. Keeping in mind that the job of most software applications is to translate inputs of one level up or down between humans and hardware, over the years different companies and developers have created different solutions to creating these applications.

Languages – like C++, Java, Javascript, Python, PHP, or Ruby – all exist to create software that will work with other software to get things done.

Whenever a company, or developer, decides to make a software application, they will choose from among these different languages. The choice will be based on what the different languages are able to do.

These languages work because they include their own software that uses them to create your application. In other words, to make Java work, you need to download the Java package to your computer, and load it up to make it work. That package will allow a developer to write programs in Java, then at the end it will package the program, or "compile" it into an application you can run on the operating system you created it for.

Some more modern languages, or versions of languages, like Python 3, can work without being compiled, but the general point is that computer languages are what you use to make an application, via their own software development environment.

And these exist in lower levels, like C++, where they talk to the hardware more directly, or higher levels, like Javascript, where they talk to another application, like a web browser.

When you are looking to get software built, or even worked on, knowing which software stack was used, and whether your potential developer is familiar with that stack, is very useful. A good programmer can use any language, but in practice there are so many existing resources, tricks for using any given language well, and so forth, that you'll want to use the right programmer for the job. This is very similar to tradespeople who get very good at certain types of building, like a hospital or data center – you learn so many specifics of what that building type needs that you are much more productive than another qualified person might be who didn't have that same experience.

In this chapter we've looked at software, from the user down to hardware. Each of the sections in this chapter will help you understand how software works and what some of the specialists you'll encounter mean.

In the next chapter, we're going to look at software as part of a network. As most of your software is SaaS, it is by definition part of a network.

# CHAPTER 3

# Software Networks

## Types of Networks

Modern software almost always exists as part of a network, often more than one. There are five types of computer networks that most users should be aware of, and understand to some degree: the internet; Wi-Fi; cellular; satellite; and API networks. Each of these works differently, does different things for you, and has different risks that are worth understanding.

### The Internet

The internet is not very new, having been created by universities for the defense department in the 1960s. The internet is the ultimate open standard, and is based on some very loose rules about how to name things, how to send things from point to point, and a system of finding things anywhere in the world.

The internet does some of this magic via internet protocol (IP), which gives hardware its "address" on the network. In fact, these are called "IP addresses," and the latest version of the internet protocol is sometimes referred to as IPv6 (it's version 6).

There is nothing inherently secure about the internet, any more than there is anything safe about a highway or side road. If you're not a software developer, there is also nothing that you can see about the internet itself, as it has no graphical interface. So the likelihood of your having to do very much with the internet directly is pretty low.

In Chapter 2, we discussed how the internet became accessible beyond software engineers in 1994 with the release of, first, the Mosaic browser, then Netscape. These software tools were based on an innovation from an English computer scientist, Tim Berners-Lee. He had created a file management system called "hypertext transfer protocol," which is what makes the web work. This is why the beginning of every full web address has http:// – it stands for that protocol, or hypertext transfer protocol.

You'll also see "https://" and the "s" stands for secure, meaning that information sent via that website is encrypted and sent via what they call a secure socket. Any web-based app you use should either have https:// at the beginning, or the developer should explain why they don't. There are other ways to secure data, but this is the most basic and common.

The internet and world wide web do not depend on any one kind of physical network. Everything works over cellular as it does over a wired connection, or Wi-Fi.

Most of the time you are sending data, getting your email, watching YouTube, or connecting to your favorite SaaS software, you are using http. In fact, even when you are using software you've downloaded, it is likely to send a message back to the publisher's servers to confirm you have a license to use the software. Almost everything we do in some way relies on the internet for data, connection, or convenience.

## Wi-Fi

Most users are accessing the web via Wi-Fi, most of the time. It is just more convenient and there are almost never any performance issues versus a wired connection. Not so many years ago, accessing the internet required an ethernet cable, and many offices still have them. Wi-Fi is so much better we sometimes forget that these cables exist.

Wi-Fi uses a specific set of radio frequencies to send packets of data based on a separate protocol known as 802.11 standard. What matters about this is that over time Wi-Fi has become standard almost everywhere, and it is not inherently secure. It is possible to secure a

Wi-Fi signal, and most companies do this at their offices. However, coffee shops, airports, and other "public" Wi-Fi do not.

It is possible that in the coming years Wi-Fi will become obsolete as bandwidth demands increase and internet of things (IoT) applications require more and more of that bandwidth. I would suggest that 5G connectivity *might* make Wi-Fi less necessary, but the pace of 5G rollout is uncertain, and the 2020 pandemic has likely slowed things down further.

In the meantime, Wi-Fi security and sometimes speed are solved by larger companies by creating what is called a virtual private network (VPN). These are not expensive anymore, and are often useful when sensitive information is being created or sent. If you run a small company and want to keep your client information secure, VPNs can be a good option.

A huge issue for construction, however, is that Wi-Fi is often tough to maintain on a reasonably sized jobsite, making many types of software less powerful, and effectively un-networked until they're able to be connected back in the office.

## Bluetooth

Bluetooth is a device to device wireless radio protocol named after a Danish king from the 10th century who united different Danish tribes – the idea being Bluetooth is supposed to unite different devices and applications. And that is a great way to view the Bluetooth standard. It is not a great way to get high bandwidth work done, which is why we still have Wi-Fi, but it is a useful way to connect small devices to larger ones. Usually we're connecting peripherals, displays, and earphones to Bluetooth. However, it can sometimes be a very convenient way to connect devices that are close to each other, and you will see this sort of thing from time to time, especially for sensors, like beacons used for indoor navigation.

Bluetooth has a low-power version, called "Bluetooth Low Energy" or "BLE," that is especially useful for peripherals that have small batteries.

The key thing to remember about Bluetooth is that it is short distance, and the fact that it is very often implemented in a way that is proprietary – so Apple devices work better with other Apple devices on Bluetooth.

## Cellular

In recent years, 4G cellular has been reasonably good in the US, with fewer coverage gaps than there were in the past. For many applications this works almost as well as a good Wi-Fi connection. Many construction sites are well covered by 4G networks, making them well networked, though there remain sites that are simply too remote – especially power plants, windmills, and many mining, oil, and gas sites.

In 2020, the talk of the next protocol was continuing, and there are a few towns and cities that have some 5G coverage. 5G is worth describing a little, because at some point in the coming years it will become useful and common enough that a construction technologist should understand what it is supposed to be able to do.

5G in the US is based on a different set of frequencies than 4G and other cellular signals. 5G itself can run on a range of frequencies, but all of them are higher frequencies than 4G, and this is because a higher frequency can carry more data, which is most of the point of 5G.

The problem is that these higher frequencies are much more easily absorbed by walls and other solids than the 4G that they replace. So exactly how the telecoms are going to roll out 5G that works everywhere still has to get worked out. We have every reason to believe it will be worked out, primarily through many more cell towers than currently used.

In addition to being able to carry up to 100 times as much data as 4G, another major benefit of 5G will be its faster response time versus 4G, otherwise known as lower latency. Latency is the time it takes for a signal to get from your device, to a service or other device, and back again. 4G is about 50 milliseconds, which is very fast. 5G will be about half that, or even faster depending on where you're receiving it.

That will mean that very high bandwidth, quick response applications like virtual reality and virtual design and construction (VDC) could be done over the internet, whereas we currently need to do these things with locally installed applications.

It is likely that 5G will take some years to fully work, as early implementations of cellular technology are usually on the low end of what the technology can theoretically provide, and later implementations see the technology mature.

### Satellite Networks

Many of the issues that plague cellular networks might be solved by the many low Earth orbit satellite networks being put up by companies like SpaceX and Amazon. These will have thousands of small satellites that are close enough to the surface of the Earth to have low latency and relatively high power. At the time of writing these are still being sent up, with SpaceX's Starlink system expected to be online by the end of 2020. Pricing and availability for use by construction companies or their software vendors have not been established yet, but it is likely 2021 will begin to see this new capability come online.

Just like cellular networks, it is likely that satellite internet technology will require some time to mature. Early implementations will effectively be "pilots," as the networks become more and more mature. However, if the aspirations of Starlink and Amazon reach fruition, they promise to transform how the world interacts with information.

## Application Programming Interfaces (APIs)

All the other networks so far are about transportation, they are essentially different types of roads. In recent years, applications have begun to share information directly with each other in very powerful ways.

There is a special way applications can talk to each other directly. Known as the application programming interface, this method of application to application communication has completely transformed how software is built, sold, and planned for. In fact, APIs form the backbone of a new kind of network, where applications are able to share data and do jobs for each other in a way that creates powerful new experiences for users.

We can think of an API as a window into the functions of a given software application, where things passed through that window are in a format that is easily understood, enabling the application to perform certain tasks, then pass information back out through the window.

For this to work, though, the way the information is passed through the window is clearly spelled out, and the application or user passing information through must use this standard way to do so, and will be prepared for data coming out according to the same standard. There is in fact a general standard, called REST (REpresentational State Transfer) that most APIs use. Because this standard is so prevalent, the cost and difficulty of creating an API has gone down dramatically, making them much more common.

Imagine an actual window, perhaps at a drive-through restaurant. For this restaurant, there's an input window, where you order. And there's an output window, where the restaurant has taken your instructions, acted on them, and produced what you need. And just like an API, there are specific things you can ask for, say a burger with fries. And a specified way to ask for them – if you said I'd like 100 grams of beef, cooked on steel, with thinly sliced cheese, between two pieces of leavened bread, it might not be understood. But if you said "Quarter Pounder with Cheese," it will be. That's what an API does – ensures that anyone can access the functions of a given application if they have the right information – and that information often includes an API key, which does what all keys do (it lets you in).

APIs are used all the time – for example, when you use your Google or Facebook account to sign into a website, that website is using their APIs to pass a request to authenticate you, and if you have an account with a cookie in your browser, you get authenticated.

In recent years, companies like Salesforce have leveraged APIs to create a platform that allows other companies to easily connect to Salesforce, pulling data out, or pushing it into Salesforce, so that users can gain functionality that is useful to them. This is very often useful when a user and their organization has specific needs or interests that the platform has not decided to develop. For example, many companies will connect their Salesforce accounts to things like accounting software, or learning software, and many others. In fact, at the time of writing Salesforce had over 3,000 third party apps on their app exchange, with over 5 million downloads.

The ability to connect one large system to a multitude of smaller, more narrow solutions has enormous power, and is already an important part of how companies like Procore go to market. In the last few years Procore has created the App Marketplace, which by early 2020 had over 220 applications that connect to Procore in a myriad of ways.

For example, in their app marketplace is a company called ArcGIS, and as their name implies they are a viewer for GIS, or geographic information systems. Their service integrates well with Procore's existing ability to display plans and drawings, by allowing these to be mapped and compared with high quality maps. In the words from the Procore website:

With the ArcGIS Viewer embedded experience you can:

- Integrate your ArcGIS Online account assets with your Procore projects
- Overlay Procore project photos on your ArcGIS Online Map View

Procore will integrate with everything from Structionsite for documentation, to Microsoft Project, Oracle Primavera Suretrak, and about ten other scheduling apps, and so on.

For at least the past decade, an ongoing concern about using construction software is that different things don't talk to each other. That is still the case for many legacy software products, but the majors like Autodesk, Procore, MCiT, and others, are all in varying stages of creating an open and easy platform for partners to connect and provide seamless value to their customers.

The way all of this works is because these platforms create an API, often a very complete one. Continuing the Procore example, they provide extensive documentation openly on their website, so that partners can begin creating an integration on their own time, then work with an internal team at Procore to test, refine, and then market the integration on their App Marketplace.

Autodesk's Construction Cloud Connect does some of the same things, as does CMiC and others. Given the success of software companies in other industries with this platform approach, and the general trend towards making APIs fundamental to doing business, you can expect that more and more construction software companies will connect to each other this way.

In summary, APIs are gateways that allow one software product to "borrow" the functionality of another software product, and in the process pass key data back and forth, to make a more complete offering for the user.

## Networks of Technology

These gateways aren't just for daily reports and design software. APIs are how everything from machinery to tools to construction management is beginning to get knitted into a jobsite-wide network of products that talk to each other. These products can be grouped into three types: tablets and phones, machines, and sensors. In a later chapter we'll dive into sensors and the internet of things more deeply; for now, what matters is that all three of these products connect via software, and that software is itself becoming more and more powerful as a result of these connections and the data they pass back and forth.

This increasingly interconnected network in turn will allow superintendents, project managers, and owners to better understand their jobsite, and manage it much more effectively. In most current construction projects, management is looking at information that is days old, which means that any issues are often dealt with well after they start to impact the job. The value of data, software, and the networks that tie them together is the bridging of that time gap between when something occurs and the team's ability to address it.

But for that to happen, an understanding by field personnel of the value data is absolutely crucial. As an industry, construction generates an incredible amount of data, but very little of it gets used. In fact, a 2018 study by KPMG, an accounting firm, showed that upwards of 95% of the data created on a jobsite is not used in management. Much of that is because the data sits on a piece of paper, but some of it is because data is not reported as completely or usefully as it could be.

## Data

As I have spoken to senior leaders across the industry about construction technology, one of the areas of concern has consistently been that folks on the ground don't always understand or appreciate the value of the data that's getting collected, or see why it's needed. After all, most of these field teams are experienced professionals who really know what they're doing.

Even more concerning is the sense that data is being "weaponized" so that general contractors have another way to put pressure on sub-contractors. We have inherited an industry where some contract terms encourage adversarial attitudes, and the reality is that disputes are common.

However, the only way a large, complex organization like a construction team can improve over time is through benchmarking and the sorts of analysis that require data to conduct. The old habit of relying on personal experience, intuition, and "gut" to make decisions has been proven less effective over and over, and is being phased out across the industry.

At the same time, data and the software that collects it provide a transparency that helps everyone. Disputes are won or lost based on documentation, and that documentation is powered by data.

So let's talk about what data is good for, how it gets used, and how to make sure what is collected is collected well and not used improperly.

## Why Data?

A superintendent, foreman, or just an experienced professional on a jobsite will be able to see how things are going on any given day. They'll be able to notice patterns in how the team works, spot issues as they arise, and act to head off those issues. That is the way it's done now, and has always been done.

But is that going to continue to be good enough? Consider how packed and almost unreasonable a construction superintendent's day is now. Many wake up as early as 4 a.m., get to the jobsite at 6 a.m. to do a walk through, and end at 6 p.m. or later, with the whole day spent fielding issues, inspectors, owners, RFIs, rework, and more. Demands for speed, safety, environmental protection, and now social distancing, keep increasing, giving the team on the ground less and less time to breathe.

And it is no easier for the trades, who need to jump from job to job to make their numbers work.

There simply is no time for most people to think about longer-term trends, to see supply chain problems that are bigger than their jobs, and so on. Every trade and general contractor (GC) I've spoken with sees the need to think about how they do what they do, to analyze what's working and what needs improvement, but none of them have the time to hold post-mortem meetings, or brainstorm about more effective ways to work.

Construction companies of all sizes seek to grow and become more efficient, and that means that management needs to understand the status of jobs, teams, and processes. And beyond just the operating of the business, without a very clear picture of how long things take to get fully completed, estimating becomes a game of sandbagging and assuming you'll make up the difference some other way.

This sense of no time to think is not unique to construction, and across industries the answer has always been the same. Collect data along the way, and use it to shorten the process of analyzing processes. In fact, by making data part of everyday operations, companies from manufacturers to designers to heavy transport have all found ways to accelerate improvement without dramatically increasing time spent meeting and doing post-mortems.

For example, do things cost what you think they cost? Do processes require the amount of man-hours you think they do? Cost of labor is the most common area where gut gets it wrong, because we tend to over-weight high profile steps like delivery, and under-weight the mundane and sometimes painful steps like rework and quality assurance.

For example, I got a chance to speak with CJ Best, Director of Manufacturing at McKinstry, a mechanical contractor in the Pacific Northwest. McKinstry is known as an innovative company with a keen focus on manufacturing as part of how they solve client problems.

CJ told me how they'd instituted daily reports and hourly tracking for their VDC, and in the process got a much more accurate picture for how much time and effort certain steps in the process, and certain types of VDC require. The point of this exercise was not to create a tool to push their people to work harder. Instead it was to feed into their estimating process to much more accurately estimate costs, margins, and risks for new jobs. In the process, CJ tells me it has dramatically changed their estimates going forward.

Data like that can be collected through software like Fieldwire, Fieldlens, and of course Procore, Autodesk, CMiC, and others. While most jobsites will have daily reports of some kind or another, they are too frequently paper, or very narrow tallying of hours. What if you could collect a little data, every day about how the team is doing and record inefficiencies and barriers to getting the work done, and so on?

## How Data Gets Used

The real value of daily reports comes when they're aggregated together and viewed alongside other data, and to do that most companies will integrate with a larger player like Autodesk or Procore. These software products come with easy analytics packages that allow you to ask simple questions and see the answer both as a number and also as a graph.

In fact, one of the most important things software can do for you is to take big piles of data and crunch them into an easy to understand, quick to interpret graphic. The power of graphed data can be overlooked, but consider this – the largest part of your brain is the visual cortex. You have more ability to spot patterns and see trouble through your eyes than in any other way. This is just as true for experts as it is for novices. So always ask for data to be graphed if you have the chance.

The point of collecting data is so that the entire operation can be run more safely and efficiently. We are better able to do this based on data, because no one person sees it all, at every level of detail, and remembers it all perfectly. It is important to separate what people are good at, and what our tools are good at. Computers have perfect memory, lightning fast search, and ability to present analysis to humans. But they have zero understanding of what any of it means.

The job of humans has always been to apply experience and judgment to the job, and that has not changed. Data does not make decisions, data does not replace the value of "gut," it is now a tool to support the intuition of experienced managers everywhere.

This opposition between technology and data-driven decision making on the one hand, and experience, judgment, and professional intuition on the other, is mistaken. Instead of having computers replace human judgment, it is precisely when you have software that can pull in data, analyze that data, and then show you results that you need a pro to make sense of it.

Nancy Novak, chief innovation officer of Compass Datacenters, tells how she has always trained her teams to view data as a starting point. Each chart and seemingly clear and conclusive data analysis needs to be gut-checked, because the data can be way off. Most of the time, the data reveal facts and trends that the team would never have seen in the normal course of business, but even then what those facts *mean* is never revealed by the software – it is up to the construction professionals to understand.

That is a general point about construction technology – it does not replace what humans can do, it augments what humans can do. And in the process, we as individuals, companies, and an industry can do things we could never do without that technology's help.

No one disputes that having a forklift is better than 10 men scuttling around trying to carry things around a jobsite. No one disputes that having a crane makes it possible to construct much higher, stronger, and generally better buildings than we could if hoists and pulleys were the only way we could lift things, not to mention the dramatic improvement in safety.

Data does the same thing, just in a less visible way. I like to make the analogy that data and software are like a construction tool and its consumables. Welders need wire, nail guns need nails, mixers need cement, and so on. In today's world, superintendents and managers need data.

But unlike any other consumable, data doesn't "run out" when you use it. You can share data internally and with other companies and it doesn't get used up, or reduce in value. In fact, because we use data to see things we cannot see with the naked eye, very often data is *more* valuable when shared.

For example, knowing your team's productivity will help you manage them, but will not help you see how well you *could* be doing. This is where benchmarking is essential – by sharing your data with another team, or group of teams across the company, you can all go beyond just managing your teams; now you have what you need to improve.

Earlier, I raised the concern that gets brought up often – that data is used as a weapon against field teams. Benchmarks could easily be an example of this, if that's where you allow the conversation to stop.

What if instead of just looking at how much gets done, data included *how* things get done? What if daily meetings included a brief discussion of processes used, issues encountered, and so on. These can then be entered into daily reports or other pre-existing software to collect them over time.

Then what looked like a punitive measure, the benchmark, becomes an indicator that some teams have innovated processes that others can learn. By combining the higher performance of other teams with details on how they achieved it, the organization overall benefits. This removes the weaponizing aspect by grounding team-to-team comparisons in real, actionable information.

This is exactly what was done in the car industry, and is one of the foundations of the Lean movement. The idea is that you never stop at just the results, but you look at the real process and keep asking questions until you get to something you can actually improve. Data is the only way you can do that at scale, repeatedly.

Of course, the other way to remove the weaponization of data is to take the time and learn about the data being collected, the system being used (like Procore), and ask better questions about it. There are plenty in the trades who are doing this now, because training is often free and doesn't take very long to get as good at this as you need to be.

To be valuable, data needs to be seen as leading to better questions and some of the answers, but it is rarely the actual answer. That takes judgment and experience.

It is also necessary that data collection be seen as part of process improvement, not contract negotiation. Data collection requires willing participation from teams in the field, and as soon as they see that data as being an effort to gain advantage, the quality and reliability of data will suffer, as it apparently has on many of today's jobsites.

## Gut-checking Data

As a professional with years of experience, the unique judgment you bring is beyond a book like this. However, we can introduce some tools and general rules about data that can help you apply your intuition and spot problematic data. In the process, you will also become a better steward of the data you and your team produce.

Let's back up and clarify what we mean by "data." The word gets used all the time to mean different things and a good first principles definition is helpful.

---

"Data is a collection of facts and observations that are used for operations and analysis."

---

There are a few key points in here.

## Data Is About Doing, or Analyzing, a Job

The first point is that any old fact doesn't qualify as "data." Facts become data when they're being used to either do a job, or understand that job. Think about what that means from the two perspectives of creating, and using data.

In the case of creating data, it is important that we understand why the data is being collected: What job or operation are we hoping will run better, or be better understood? In most instances this will be easy to figure out, as data are collected in the process of doing a job. We ask the question explicitly so we can clearly think about what data is important, and what data will not help.

Most data you will be personally creating is reporting on progress, quality, and manpower. Each of these can be fudged, can be reported in an unclear way, or can even be missed on occasion. At the time, it can seem like no big deal to miss or fudge the day's data, but keep in mind you are creating something that will help you and your team in time. That data will be gathered over time and help you run your jobs better, defend against certain claims, and generally make it clear to everyone how things are really going, day in and day out. To do any of this, the data needs to be analyzed as a large collection over time, and that analysis critically depends on what you and others do every day.

Most companies have faced an issue of data consistency at some point. In fact, at present most companies are *still* dealing with this. Because construction is by definition a fragmented process, with different crews on different jobs in different cities with different PMs, past data is usually in the format the PM required, not a company-wide standard.

As companies have adopted modern project management software, like Autodesk, CMiC, or Procore, the value of seeing things across the company has become more clear. These networks of software help everyone when data are clear, consistent, and on time.

The value of this for helping teams do their jobs isn't always clear, but when that data, and the way it is entered, becomes consistent, the ways it will help becomes immediately obvious.

At Harkins Builders, a regional general contractor based in Wayne, Pennsylvania, Patrick Hennessy worked with internal teams to make the way reports were filed, issues classified, and generally how data was entered consistent.

When Patrick first joined Harkins, he found that processes and the data they produced differed from team to team. Pat and his team found that the best way to tackle this problem was to form a dedicated team to go to individual software users within their organization, and create a common way to enter information and thus create an entirely new way to look at the business.

With senior level support, Pat and his team were able to bring many of these different processes into line, with the result being that management at different levels now have a powerful tool to help teams be successful both in real time, and in better preparing for new jobs.

One of the ways that Harkins ensures that their partners in the trades understand that data and the dashboards it creates are being used for mutual benefit, not just as a tool to put pressure on the trades, is that they put their own teams in the same charts and analyses as their trade partners.

This simple tactic speaks volumes about their general approach – everyone appreciates being in the "same boat," a feeling that is not always prevalent in GC-trade relationships.

Pat shared a few ways that this data allows them to manage jobs and processes more effectively, and a great example is their project risk report.

Harkins now has a report that includes 10 factors indicating that a project is at risk – things like percentage complete against plan, and other KPIs. Prior to their harmonizing the data across the company, management would only know a job was in trouble when it was likely to fail. Now the CEO is able to view a dashboard of over 50 jobs at once, and see in real time where resources are needed to help teams that are falling behind. The point is not to go and crack the whip, it is to make sure the project is a success.

Perhaps most importantly, the team has been able to correlate these 10 KPIs to jobs that are loss making. In Pat's words, "We always knew when jobs were failing, now we know why they are failing." Now they can act in time to save these jobs.

Which is our second point: Data is for understanding and analysis. Beyond just getting the job done, actually understanding what is driving success or failure has always been important. Data allows managers to see what these success and failure factors are in ways that are simply impossible by human observation alone.

## Analysis: Data + Judgment

No software can hope to do some things that humans take for granted. Software cannot understand the context of a situation, cannot intuit when things are not quite right. We are used to seeing software excel at tasks where humans struggle. The opposite happens much more frequently – things like judgment, complex understanding, and intuition are things humans do effortlessly. Software cannot handle anything beyond the narrow cases it was built for.

Analyzing data is about bringing these two complementary strengths together. Software, and the data you feed into it, can show patterns across time, across jobs, and across teams, that have been invisible to the unassisted manager. But those patterns do not automatically have any meaning, they might just be some random pattern in the data, or might be caused by something irrelevant, like the weather.

In the Harkins example, no one person had the ability to understand what's going on across 50 jobs, without software and consistently formatted data to provide it to them. Once the data had been collected, though, the viewer of that data had to apply their own judgment to decide when the data really showed a job was in trouble, and when it did not.

Because the manager can get to know team leaders personally, he or she can dig into the situation and find out if there's anything going on beyond what the data shows. That same manager, having been involved in construction for years, will also be able to interpret the data in the context of other facts and a general sense of how jobs succeed or fail.

The power of modern construction technologists comes from this application of hard-won construction experience to newly created data and analytics that allow them to see beyond just their own job or immediate group.

### Data Standards

APIs create a fantastic way to knit different software systems into one network of data and useful functions. However, an API is only a window, a conduit between different software products. In some cases, the data that one software product creates is useful for another product. But in most cases, the API of one product requires that others conform to its specific requirements to be useful, and most software products only do this with a handful of other products.

The arcGIS example from earlier illustrates my point. arcGIS uses a standard format for its geographic information. This is a format that Procore was already using, because it is a standard. The same is true for many other file formats, like a JPEG, or an MP4 – these standard formats mean that you can easily share photos and videos. However, data standards need to be about more than file formats, because we do more with data than just display media.

The point of data is to group it together, to be able to run some analyses against it, like averages and percentage of change, and so forth. To add things up, they need to be in the same format, and this almost never happens naturally.

Recall earlier when we discussed Harkins builders. They had to have a senior level team spend almost two years to get their different project teams to use the same KPIs, use the same processes to enter them into the system. They spent money and resources to change how they handle data.

Now imagine you have 50 project management software companies. Or you have a design software, a daily report software, a safety management software, and so on. Each of these companies will have built their products independently, and will have made decisions about how to format their data.

As a result, software from different companies often doesn't work with each other, especially when it comes to highly detailed data about job performance. This causes companies, especially specialty trade contractors, to double and even triple enter the same data into different systems.

Across my time in construction technology, this has been the most common complaint.

Two things can be done about this. The first of them is happening pretty quickly, and that is the expansion of the big software platforms, like Autodesk Construction Cloud, and Procore. By bringing more and more functions into one platform, issues of data standards become irrelevant. However, no one company can be the best at everything, especially in software where new ideas are popping up all the time.

The second thing that can be done to encourage data standardization is what happens in many other industries – someone creates a standard. In the broader software world, that is often the IEEE (Institute for Electrical and Electronics Engineers) – they're the group who set the Wi-Fi standard, for example.

Because construction is so fragmented, but also too narrow for a group like the IEEE, it has been difficult to evolve data standards that are broadly adopted.

There is one emerging standard that is the brainchild of some of the original construction tech enthusiasts. Called the CDX, or construction data exchange, this standard has been painstakingly created by a nonprofit group called the Construction Progress Coalition.

The CDX is as much about the approach toward data standardization as it is an actual standard, because Nathan Wood and his team recognized early on the same lesson that Patrick learned at Harkins Builders – the industry is loaded with highly experienced, deeply professional builders who rely on what has worked in the past to ensure they successfully complete the next project. That experience usually leads to a reluctance to try new things that might cause complexity and room for error in the future. And those same people are too busy to carve out time to understand a new way to enter data, or file an RFI, submittal, or other paperwork.

The CDX is currently working hard to create a new RFI standard, and their success in getting people to the table shows how important this is, but the fact that they have not yet had final success in getting a standard broadly adopted shows how hard it will be for external standards to ever be adopted by individual firms, especially software firms who are very convinced their mousetraps are better.

Of course, BIM is the ultimate software standard, but there really isn't a "standard" so much as a collection of products and practices that fall under that name. If something as fundamental as BIM has taken decades to become a standard, we can understand how resistant to other standards the industry remains.

## Software as a System

In this chapter, we've gone from the ground up to understand the basis of networks, how they work, the all-important API, and how data can be used to improve the business of pulling buildings out of the ground.

The issue of understanding data, and how to analyze it and use it will continue to be important for anyone looking to master the construction technologies of today, and tomorrow. We covered it here at a high level, but there are many free resources out there that offer a deeper dive, not least of which are my courses on procore.org.

# CHAPTER 4

# Construction Software

According to data from PMI-Plangrid, "35% of construction professionals' time is spent (over 14 hours per week) on non-productive activities including looking for project information, conflict resolution and dealing with mistakes and rework." This 2018 study found that these non-optimal activities cost the US construction industry over $177 billion in labor costs alone.

Software for field management is absolutely essential to the continued growth of the construction industry. More than any other area of construction technology, field software has seen a boom in innovation and product options for field operators, and that has also led to "app fatigue." In this chapter we'll review what's out there and draw some lessons for thinking about field software, as it will remain one of the most important areas of construction technology development.

## Beginnings

It is a common complaint amongst field teams and construction technologists that early field software wasn't very helpful. Much of it began as accounting software that was extended into the field to capture accounting data. And much of it had the same stiff user experience as accounting software, didn't integrate well with other software, and generally left a bad impression. Most of all, early software required changes in daily processes that didn't seem designed for the field, but were to the benefit of office teams.

This idea that software can impose new processes on established teams is an old one. In the 1990s, the manufacturing sector was revolutionized by the introduction of "Enterprise Resource Planning" (ERP) software, spearheaded by a German company called SAP. ERP products from companies like SAP, Oracle, and others were in fact world-class software that did genuinely useful things; indeed, they formed the heart of digital transformation for many companies. But implementing them was much more disruptive than is today necessary, for a few reasons.

Traditional ERP was like all software in the 90s – something you installed locally on your computer or company servers. As a result, there wasn't any way for SAP to provide ongoing improvements; in fact, each release was a big deal and required armies of consultants to implement. So, ERP implementations had to be inflexible, one-size fits all, and as a result required the user to conform to the software, not the other way around.

As we discussed in Chapter 2, Software as a Service changed this paradigm so that software now is much more flexible, constantly changing, and configurable to what the user needs. And with that flexibility has come a more important change – software companies now compete on user experience, on making the software easy to use.

Today, ERP systems still exist, and are often sold on a SaaS basis, and no longer expect that they can demand change from their users.

## The Problems with Field Software

Almost everyone resists changes they don't see as necessary. And absolutely everyone resists changes that needlessly complicate their lives, and unfortunately for many in the field that has been their experience with technology generally, and software specifically. It is easy to point to earlier, accounting-led software as the excuse for this, but there is more to it. The reality is that most software is not designed for use in the field, because it is not developed *by* people in the field.

The software development process starts because someone has a need, will pay for it, and is present when that software is being developed to ensure it works for them. Looked at that way, is it any surprise that software started with people who sit in offices, using other software all day long? So, design software has a huge head start over anything related to the field. And accounting software was the first to be introduced to the field, for the same reason.

The people who design and build software will always be biased towards their own habits where they work. But it is the job of software vendors to go find out where their products will be used, and ask if they are really helping the crews in the field, or just helping management. There is no reason both cannot be true, and the imbalance between easy-to-access office teams and critical-to-project-success field crews is starting to be addressed. The voice of the field is starting to be heard, driven by smart innovators, unions, and field teams willing to spend the time to teach technologists what will really work on a jobsite.

Jake Olsen, CEO of DADO, a construction software company, illustrates this when he describes how his company came to be, and how they still develop features.

DADO started as part of a technology incubator program funded by DEWALT, based in Silicon Valley. Jake and the team originally came with a BIM-related idea that they were 100% sure was going to be a success, and they presented this idea to their mentors at the incubator. In three questions they were sent back to the drawing board:

"Who is this for?"
"Do they need it?"
"What do they *really* need?"

The DADO team spent over three months interviewing field crews, from apprentices to foremen; in addition to just about every other part of the construction value chain. What they found was that the biggest obstacle was finding the right thing at the right time, quickly and easily. The goal they landed on was to remove barriers to accessing the newly digitized information, and in time they settled on voice as the main way to do this, in addition to web and app-based interfaces.

The conversation with the field didn't stop there – in fact, DADO regularly meets with trades professionals for happy hours, lunches, and other forums that allow them to test new ideas and ensure that they're creating something that works for the field.

Jake points out that lots of software that seems to increase productivity only does so for one side of the information equation, usually the office side. The reason, again, is office workers are easy to find, already use software all day long, and immediately see the value in using it. But it is not obvious that a field crew's productivity is enhanced by stopping what they're doing, taking their gloves off, and punching their daily reports into an app, versus just writing it down on a piece of paper. The benefits for the company overall are obvious, and of course it is worth converting that sort of information digitally – but is the input mechanism optimized for the field worker?

DADO and others have looked at the problem from the other perspective, asking how the user experience in the field can be optimized to make sure that data collection doesn't just push complexity and inefficiency down to the field.

The reality is that most smaller software companies can barely get the product out the door, so they rarely spend the time to go learn from less-accessible users, like the trades. The irony is that many startups are so busy building what they think is needed, they don't have time to go ask what actually is needed.

We're seeing the situation change, starting with the big construction management software companies that have begun to really cater to and court the speciality trades. At the same time, the unions and associations that represent trades, like the United Association (UA), National Electrical Contractors Association (NECA), Mechanical Contractors Association of America (MCAA), and the Association of Union Constructors (TAUC) amongst others, have been investing serious time and money educating their members on different software products, and providing input for software companies to begin to bridge the gap.

The key takeaway for readers in the trades, though, is that the providers of technology are generally eager to make things you will use, and your input will be much more valued than you might think it is. If you'd like to learn more, just reach out to bluecollarlabs .com. They partner with tons of construction professionals and the software startups working to create products for them, and would love to hear from you.

## Overview of Field Software

To understand software in the field, let's look at what's out there from large to small, and in the process understand how emerging platforms are likely to make adopting multiple software products easier. We begin with the big software platforms that promise to bring the entire construction process under one roof – construction management software.

## Construction Management (CM) Software

There are a number of companies who provide software that will manage construction operations to one degree or another. Trimble, Autodesk, Oracle, and Procore all offer platforms that will do everything from daily reports to scheduling to BIM management and much more.

CM software companies are in the middle of a fiercely competitive battle for market share, and are all introducing new features and entirely new functions constantly, so we'll leave product descriptions to their websites, and focus instead on why you should care about CM software, what to expect of it, and how to make it work for your operations.

So what does CM software accomplish? Other than turning paper into screens, and making it easier to find things, what is the point? To understand what CM software is there for, what it strives to become, and what you need to know about how it'll change in the future, let's step back and consider what goes on in the field.

Every year, over five million men and women from over 26 recognized trades, and at least 730,000 contractors of various sizes and specialities, come together to create thousands of structures. Each job is unique, each jobsite is a carefully choreographed, or chaotic, dance of work being put in place as efficiently and safely as possible.

Danielle Dy Buncio, Founder and CEO of ViaTechnik, a BIM and construction technology consultancy, summed it up: "Construction has some of the same challenges as any production process would, but we have some truly unique challenges as well. General Contractors have to bring together, at the right time, a multitude of specialists who may be managing their own supply chains, and who may not be under the GC's contractual control. All while basic factors like weather can shut down everything unexpectedly."

The complexity that this implies is potentially endless, and in practice means that managing these potential risks is a task for everyone on a jobsite, from the pipefitter making sure he's got extra gloves and welding supplies, to the superintendents who begin every day walking the site to check that work is put in place, and make sure there are no surprises.

Construction has been described as being a game of pushing off risk onto other parties, a description that is borne out by many construction contracts. The uncertainty of construction makes this risk offloading a natural impulse, because no matter how much risk you contract away, there is always more.

Underlying this risk management is that key term: uncertainty. Not being able to count on something is expensive. Uncertainty about productivity means you need to hire extra workers. Uncertainty about weather means you need to schedule extra days. Uncertainty about who is doing what means you burn man-hours in extra meetings. I come back to that PMI study about $177 billion in wasted hours looking for information – that is pure uncertainty, at its most expensive.

In fact, uncertainty is how insurance companies and Wall Street put a price on risk, because it can often be measured. The finance industry measures uncertainty by how much variability there is within a process, and we know from another study that the average crew productivity can vary by as much as 100% from day to day, and different crews doing a similar job can vary by as much as 500%. Construction has a great deal of uncertainty, and therefore a great deal of risk.

Uncertainty is what drives productivity down, it is what wastes the time of everyone from the CFO to the framing contractor because they don't quite know where the job is, or whether the next two weeks will look like the plan, and so on.

## CM Software's Job #1: Reduce Uncertainty

The first thing CM software needs to do is reduce uncertainty. Unlike any other solution, modern CM software is able to reduce different kinds of uncertainty at all levels of the organization. Let's look at the three components of uncertainty for a given decision maker, and how CM software addresses each: good information, good analysis, and good presentation.

## Good Information

A construction site is a hive of activity, and each of those activities will change some aspect of the project, and thus the information about the project is constantly changing. Combine that with weather, government, and regulatory impacts, and changes that come from the design and owner side, and it becomes clear that a major source of uncertainty is just being up to date. In the past, different groups would get information at different times, so there would naturally be concern over whose version of the "truth" was the right one. The effect on productivity of this sort of uncertainty can be significant.

Products like Procore, PlanGrid, Viewpoint, and other CM software go to great lengths to connect as much information to the user in as many ways as possible. While this doesn't guarantee the underlying quality of the information, it does mean that everyone on the jobsite can have access to the latest drawings, notes, and schedules at any time.

The term "single source of truth" is often used in construction, and that is a primary role for CM software, especially in the field.

Schedules, drawings, and reports are all pulled into one place. Uploading new drawings, and keeping track of versions is fast and automatic, because everything is stored in one place on the cloud. This is of course true for all of the major CM software providers – Procore is notable for how easy they make it to get trained on their software for free.

So how do we do this – how do we get all that information into a construction management software system? This is what we mean when we talk about digital transformation.

The first "phase" of the digital transformation of construction is to turn all of the paper workflows, and as many of the ad hoc conversations as possible, into a digital form that allows for instant collaboration and real-time updating of everyone who needs it. This first phase is very far from over, but the key technologies are here and have been worked out to the point where they're ready for any level of company and worker to adopt.

What's slowing adoption? Why isn't everyone using CM software? Companies and individuals adopt a new technology because the value it provides is greater than the cost. But the cost potential adopters consider isn't just the day-to-day cost of using it, it is the mental cost of changing processes and training on the new software. It is often as simple as teams and workers not wanting yet another thing to think about, when the software or paper they've been using for years isn't "broke."

The question of enterprise-level cost versus value has also been answered a dozen ways, from calculating superintendent time saved finding the right document, to reduction of mistakes and rework, to faster RFIs and beyond. Just about everyone who has adopted CM software recognizes that it is a huge improvement – it is worth it.

The remaining issue is now the perceived cost, the mental hurdle of "do we really need to do this now?" And increasingly the answer is yes.

Good information isn't just about managing information, it is also about producing data so that the CM software can do its job. Construction produces enormous amounts of data, most of it in a form that's not as useful as it needs to be. This first phase of construction digital transformation has included companies realizing that they have to create internal standards for how information is entered into their CM software – a problem that didn't really exist before but is crucial to getting value from CM software.

For example, the way RFIs are recorded and processed can vary by superintendent, project, and over time. Everyone knows RFI processes are one of the biggest pain points in construction, and any attempt to fix this overall issue needs to be able to look at data in aggregate. As a result, companies across the construction industry are standardizing how RFIs are entered, and thus empowering their teams to track and analyze the issue more effectively.

Then there is the issue of connecting systems, as we saw in Chapter 3. Many companies, especially sub-trade contractors, report needing to double enter information, especially daily reports, because their chosen system doesn't work with their upstream clients, usually the general contractor. This issue is unlikely to go away. Many large business systems do not directly connect to each other, requiring expensive development of bridging software to translate or connect them.

Indeed, many companies in the construction industry have hired consultants like Jonathan Marsh, of Steel Toe consulting, to help integrate different systems and more generally plan and manage their digital transformation.

Overall, we have three ways that information from the field is getting connected together:

1. It is all coming from one system, like Procore or Autodesk Construction Cloud
2. Companies are building their own solutions, often stringing together smaller software products
3. Standards are developed that allow all software to pass information to each other seamlessly

None of these approaches is working completely right now, most companies do a blend of the first and second, and rely on the third for file types like Revit or OBJ files. The history of data connections in other industries suggests that this could continue for a long time, unless a big customer or group of customers demands a standard approach. This is not likely, because no one company has the market dominance to make such a demand stick.

This doesn't mean these initiatives are doomed to failure, just that some systems, and some kinds of data, will be easier to connect than others. In an industry with 730,000 companies, this sort of segmented market is almost inevitable anyway.

What is probably most likely is the rise of "API networks," where smaller companies connect into the platforms of larger ones, like Procore or Autodesk. That's happening now and might wind up making data interconnection much less of an issue in the future.

### Good Analysis

Just having data from different sources in one place is already a big plus, but CM software comes with a full suite of visualization and other analysis tools that help tell a story from that data. For example, knowing a job is ahead of schedule is great, but knowing why a job is ahead of schedule is how crews improve over time, and how the organization begins to drive higher productivity.

Kris Lengieza, director of business development, is leader of Procore's marketplace, where data from third parties is integrated into the

Procore system. Prior to Procore, Kris was head of operational excellence at Stiles construction. Kris tells the story of how he used their CM software to analyze RFIs, job profitability, and job size. Kris found over these jobs that project managers who issued more RFIs tended to run more profitable projects. What was important, though, is that the profitability came from better documentation, not the RFIs themselves. By carefully documenting RFIs and the change orders they produced, the company was more likely to get paid for its change orders.

Now imagine two scenarios. In scenario #1, management walks up to a project manager who is a 20-year construction veteran and tells him or her that they aren't documenting their projects well enough. In this scenario, there are only a few examples to prove the point, and the PM argues back that those are just exceptions, that most of the time it's not an issue.

In scenario #2, management pulls all of the PMs together and presents a comprehensive analysis of 50 jobs, showing that higher documentation leads to a higher percentage of paid change orders, and that this leads to better margins for those jobs with higher documentation. Now imagine that management has calculated the average bonus this margin will translate into for a PM.

Construction management software should be the difference between these scenarios, and many others like them.

A word of caution when it comes to this sort of multi-sourced analysis. Just because an application has produced a number, or an output, doesn't mean it is valid. Always, always apply your instincts and natural skepticism to information that comes from software. Most of the time this information is of high quality, and will dramatically improve your ability to understand and thus manage projects. However, there is no guarantee that all of the data put in was done so correctly, or that it is complete, or that there is enough data to draw conclusions. We as humans get a little lazy when we see information in printed or digital form, we assume it is valid. Best to double check.

As a quick example, whenever you see a percentage in a graph or report, it is useful to ask how many data points that percentage represents. Graphs are notorious for fooling people with different scales, or presenting pie graphs based on a really small number of data points.

For example, if you read that 17.5% of crews show up late, it is worth asking how many crews they surveyed or otherwise collected data from. If the answer is 40 crews total, that means they are saying 7 crews were late. Now that's not great, but is the 0.5% meaningful? In fact, is the difference between 17%, 17.5%, or 20% that meaningful? Each of these percentages, when taken as a percentage of only 40, is still about 7 ($40 \times 0.17 = 6.8$; $40 \times 0.175 = 7$; $40 \times 0.18 = 7.2$).

This is often referred to as the difference between being precise, and being accurate. And if analysis is more precise than it is accurate, it can be misleading. You will find this too often, and it is why you should look at graphs of data at least as much as you look at numbers – you'll pick out oddities more quickly, and graphs can get you past the sometimes distracting decimals to see the simple picture. But always ask how many data points were analyzed to get to a given analysis.

### Good Presentation

Having good information and analysis is important, but so is being able to access that information. Since most software is accessible on users' phones as well as laptops, it should be a given that you can do this.

What really makes the difference is user experience, and user interface. Can the user do whatever they want, whenever they need it? Apart from different permission levels, where a sub-contractor might not get to view scheduling or other information, the user should never feel like they don't know how to do what they need to do next.

A simple rule of thumb for field software: it should feel invisible. It should be so easy and intuitive to use that users forget they're using one or the other product, and just do their jobs quickly and easily.

The thing about "intuitive" with software is that there is nothing naturally intuitive about tapping a glass screen and looking at computer-generated models on a dusty jobsite. Intuitive really means it works like software we've seen before, that it uses common website and computer software design choices. Most CM software products do; some better than others.

The other thing about being intuitive, though, is that at every point there is an easy menu that gets you to the next thing you might want to do. And here some software products are definitely better than others.

UX and UI will be constantly changing, and it is part of why testing software with real field personnel is always critical. It is also very often the case that the UX can be adjusted to work for the user's specific workflow. Configuring software to work for exactly what you need is common, and a process that should be budgeted for if CM software, or any software for that matter, is to be rolled out at scale. As we discussed in Chapter 2, as an individual it is worth the effort to learn how your software presents data and options – if you master their user experience, software can become an incredibly valuable tool.

## CM Software as Platform

Construction management software has started to become the "operating system" for the field, and will continue to do so in the coming years. Part of how it will do this is by accepting data and functions from other applications via their APIs. The big CM platforms all have special programs that make selecting and buying these additional apps easy, with integration one or two clicks away. Procore has their App marketplace, Autodesk has Autodesk Construction Connect, Trimble and CMiC both have dedicated integration programs that help providers integrate, and users can connect these apps to their core services.

The innovation that platforms like these can unleash is remarkable, as the entire startup ecosystem, with hundreds of developers, small firms, and innovative product designers can now add to one user experience. Salesforce has proven this model in the sales and marketing field, and I believe we will see the same thing in construction.

## CM Software's #2 Job: Improve Productivity

Risk and uncertainty are important, but so is productivity. It's worth clarifying what we mean when we say "productivity," because most people don't spend very much time thinking about their activities in terms of value per hour, but that's what it boils down to. Productivity is measured as dollars of value produced in one hour, averaged across crews, companies, and industries. As an individual manager, you might count units of work done, but to analyze large groups of workers, we use dollar value.

Bureau of Labor Statistics data has been quoted widely as showing construction's low productivity growth. Let's take that apart, so
we can understand what frontline workers and managers can actually
*do* about productivity, and how software helps.

There are two kinds of activities on any project. Those that add
monetary value, and those that do not. Since productivity is measured
by dollars, not any other source of value (like safety or being on time),
anything done by the company that does not directly result in something that can be paid for is not adding value used for productivity
calculations.

To understand productivity, we can assume there are three kinds
of activity that happen in construction: The first category is activities
that directly result in getting paid, which in the case of construction is
*only* putting work in place. The second category would be activities
that directly support putting work in place, like safety inspections,
planning, and so on. The third category consists of the many things
that take up time but don't move the project forward, like rework,
and coordination costs like waiting for one crew to finish so the next
crew can start.

The goal of any project is to spend the largest percentage of time
possible directly putting work in place, driving time and cost out of
support activities, and eliminating rework and coordination costs. CM
software exists to make each of these things happen. Let's review the
two main tasks of reducing support and eliminating waste, through
the lens of how you, the operator on the ground, can take advantage
of what the software has to offer:

1. **Driving time and cost out of support activities**: Complex
   projects require significant time spent just keeping track of
   what's going on. CM software helps in four ways:
   a. *Managing planning*: there is now one place, automatically
      updated immediately, where plans exist. This system records,
      in real time, work done towards the plan, events that interrupt
      the plan (like an RFI), changes to the plan from the design
      team, and so on.
      *What you should know and do*: These systems usually don't
      require that teams change anything about their processes,
      except that they need to enter information consistently.
      Make sure to standardize terms used, classifications, and

anything else that can be standardized across teams, so that managers and analysts can leverage the system for higher level reports.

b. *Managing drawings*: Everyone is ultimately responsible for delivering against the approved drawings. Finding the right drawing and ensuring it actually is the current drawing, was an expensive process in the past. Modern software ensures that the most up-to-date drawings are available at all times, on a laptop, tablet, or phone.

*What you should know and do*: Take the time to understand exactly how to operate the software – this is designed to be easy, but will be worth the effort to get a little training via online tutorials or just asking a co-worker.

c. *Managing actual work:* Daily reports are the single most important ongoing source of information for project managers and teams, because they allow mundane, ongoing work to be understood without the filtering of memory and recent events. No one remembers the minor events of three months ago, but by keeping good daily reports these can be readily recalled as needed. And of course, if there's ever a dispute, the best documentation wins.

More broadly, over time we are going to find that the information from daily reports, as well as other routine data inputs, will be used to create new process improvement recommendations.

*What you should know and do:* Push your crews to enter information as consistently as possible, including incidents and sources of process friction.

d. *Coordination*: Drawings, plans, hours, safety, and inspections all require that many different teams, some of whom are not contractually connected to each other, work together and minimize wasted time. CM software is a central source of truth that can be leveraged to minimize coordination effort and time.

*What you should know and do*: Learn to use the scheduling and notifications features of CM software, they can make keeping track of crucial inspection dates and other time-critical events much easier.

2. **Eliminating waste**: Many of the same software functions that will lead to more efficient support of work will also lead to at least the partial elimination of waste. The reality, of course, is there will always be time spent on non-productive activities, but if we can leverage the ease of document search, the single source of the most updated drawings, and immediate access to the most updated information, time spent searching for information, waiting for document retrieval, and other wasteful activities can be minimized.

## CM Software Job #3: Process Improvement

Digitizing paper workflows is usually thought of as the first phase of construction's digital transformation. Just converting paper to software creates all of the value described earlier. However, that's not really transforming the business, just making the paperwork faster.

The true value starts to come from using software and data to analyze how things are operating and find improvements. Process improvements come about because someone is willing to spend the time and effort to convince all those using the process that there is a better way.

In an industry where many have been doing their jobs for decades, and can point to massive buildings to show that the old way isn't broken, how do you get people to do things a new way?

The answer is to prove you are right. The only way to prove that a new process is better is through data. In Chapter 1, we discuss the difference in acceptance between technology you can see, and technology you cannot. This same gap applies to effects you can see with your own eyes, and effects that require technology to see. It is easy to see that giving everyone access to the same information at the same time is going to help. Not so obvious is how analyzing data can expose some assumptions as being untrue, or show you that the team spends more time on some parts of the building process than you thought.

Using technology to see things that you cannot see with the naked eye is not new. Electricians base their entire profession on using technology to see amplitude, voltage, frequencies, and so on. Everyone relies on lasers to measure jobsite dimensions and grading, and you can't see either of these.

CM software, for the first time, allows managers in the field, as well as in the office, to see how processes are doing at levels that are simply not possible to see in person. Decisions can be made based on *data*, not experience or intuition. And that is how you prove you are right, versus just louder.

Todd Mustard, VP of Industry Innovation at The Association of Union Constructors (TAUC) immediately went to baseball when we discussed this. By now, everyone knows the story of the Oakland A's and how they used data on players and teams to understand things that instinct alone just couldn't handle, and in the process they created a winning team.

This is what CM software can provide, but to do so it requires that data be entered consistently, correctly, and then analyzed well.

Imagine if you took all the reams of data that are produced by construction and made it available for analysis? History and other industry experience says that almost everything about the job gets better, from efficiency to safety to simple things like comfort level.

Here are three examples of how companies have taken the data provided by CM software, analyzed that data, and made important decisions that dramatically affect their operations:

1. **Estimators**: In Australia, Scott Polson of Benmax recounts how they were able to use their CM software to analyze what jobs made money, and which didn't. In the past, they might take jobs because they knew a developer or owner, but they found that some of those relationships weren't in fact profitable. Benmax can now boast of a 60% win rate on bids.
2. **Estimating** (again): CJ Best's example of using daily reports to really understand time spent on certain VDC processes changed their understanding of costs, and led to a review of job estimating.
3. **RFIs**: Kris Lengieza's Stiles construction example shows the value of analysis; it also shows how Stiles were able to identify architects/engineers who were slow to respond to RFIs, giving the Stiles team the ability to focus on getting answers from them more quickly.

## CM Software Job #4: Single Source of Truth, Anywhere

CM software allows everyone involved in a job to see the latest drawings, reports, analyses, and whatever other job-relevant information they might need, subject to role-based permissions. This was already important prior to the 2020 pandemic, but as the lockdowns have turned into social distancing and other measures to mitigate viral spread, many assumptions about the need to do things in person have been proven less accurate than assumed. We simply don't always need to be together to look at a drawing revision and discuss it.

The fact that cloud-based technologies like CM software allow for real-time updating, instantaneous calling up of drawings and other information made the switch to virtual teams almost immediate.

Without using this sort of software, teams will store drawings and information in emails, or cloud storage (e.g., Box or Dropbox), which creates issues of who can access, tracking versions, and of course the delay between marking something up and getting it into the right folder. The impact on rework, already a big benefit, will be much more profound when fewer managers and design team members will be able to walk the site regularly.

As this book went to press, all of the construction management software providers have been busy rolling out features that support remote teams, and in time this is likely to develop into feature sets that support whatever hybrid working arrangements emerge in the years after the pandemic and its lockdowns. We can expect some return to pre-pandemic workflows but also expect that, now that firms have realized that productivity doesn't evaporate in remote teams, some level of remote working will remain.

## Field Software Beyond CM

There is no shortage of software built to address smaller pieces of the construction process – too many types and companies to include here. However, most of the decisions that you are likely to make will be about smaller solutions, not about larger CM software.

Three things you should understand and consider about any new software you're looking to adopt: Does it work for all of your users; does it work with your CM software; do you have the team to manage it?

## Does it Work for All of Your Users?

It is commonplace that only one part of the team is consulted when new field software is purchased. As studies by Dodge Analytics and others have shown, in the past it was not uncommon for field software to be purchased without testing it with field personnel.

This is astonishing.

As we'll learn in the chapter on innovation and adoption, many general contractors have created innovation groups, and almost all of them went through a similar process. In year one, they bought a bunch of tech, which no one used. In year two, they started asking crews what they wanted. In year three, they'd all set up formal processes that worked with field teams to understand needs and match them to software and other technology they'd sourced in the marketplace.

It is easy for the top 100 general contracting firms to have teams like this, and many do. However, in an industry that has over 650,000 companies below 250 employees, most cannot afford to hire a team to manage innovation. Lacking an innovation group should not stop you from benefiting from new technology, but you still need a process to ensure your investment in new technology pays off.

For firms without an innovation team, a simple four-step process can help you decide what technologies are worth trying out. These create a kind of "funnel," where each step should disqualify more and more options, until you've either knocked them all out, or landed on a few. The assumption is if you're looking, you already have identified a question or problem – something you think can be done better. So we start there.

**Step 1: Define the problem**. Give your team 90 minutes to really chart out what the problem is. Get field crews in, and ask questions like:

- What isn't working?
- How do we measure the work that'll be affected?
- Who does the work in question?
- Who would operate the software? Is that more than one group?

– What do they do now?
– *Does the software you're looking at line up with the above?*

**Step 2: Get a demo**. A live demo if possible. What's important here is to understand most software companies are terrible at demonstrating their products – they'll show you the shiny parts they're proud of, but often miss what you care about. So tell them what you care about, and tell them you want your teams to be able to try the product in the demo if possible. For many software products this is just downloading something and getting a temporary username.

The software vendor might resist this, but no one on your team can be expected to really understand the software if they don't touch it. You'll nod your heads and *mostly* understand, but nothing beats using the thing, even if it's just for 20 minutes.

This is also a good BS detector, because many, many early stage software companies will tell you that the software can do things it can't, at least not yet. It will also be a good way to see how confident they are that the software will be bug free.

It is also a fantastic way to keep your teams engaged in technology, and feeling like they're part of the decision process. Worst case, it's a training expense. Best case, you're getting an early understanding of what a rollout might require.

Before this meeting, spend 15 minutes thinking about what it'll take for them to win, and tell them that so they speak to it and you don't waste time on things you don't need.

**Step 3: Get a pilot**. You'll need to pay for this, possibly a little more than they officially charge. It is worth the money to ensure you get the time and support you need. The pilot might require that the software company do some configuration and setup for you, and you should require that they also provide training for your pilot crew.

Hopefully you were able to get a hands-on demo, so you walk into a pilot with some idea of what works and what you're not sure about. The pilot should be 100% about value delivered to the project, minus effort required to make the software work.

The software vendor should have a helpdesk or other easy support, and a sales team you can meet with routinely to make sure you're getting all the value out of the software you can. They will suggest three months, but you should feel comfortable asking for six months if you feel that'll be necessary, as many contractors do.

Be sure to ask yourself:

– Is this working flawlessly?
– Is this saving time overall? Is it shifting pain from office to field, or vice versa?
– Is this saving enough to be worth the extra complexity and cost?

**Step 4: Rollout.** Assuming it all goes well, you'll want to roll this out amongst your teams. By now a few people on your team know the product well, so can be its champions. But you should also rely on the software company for support in rolling it out – remember they make more if you buy more.

Most software companies will have a group called "customer success," and their job is to work with customers to encourage and track internal adoption of the product. They'll often do webinars, take field team calls, and provide materials that help you as the process rolls along.

One of the key things that gets missed in early rollouts is how to enter data, or otherwise use the software so that management can look at how everyone is using the product all at once. Consistency of data and content produced by software is very often one of the big value-adds that software can provide, so get the customer success team to help you think this through.

## Does it Work with Your CM Software?

It used to be the case that big software products like Autodesk, Trimble, and CMiC only worked with their own sister software products. In fact, this was true across the technology industry, as Microsoft, IBM, Apple, and other major players locked down their systems for decades.

However, the advent of APIs and cloud computing have led to an opening of systems, with the big software vendors increasingly becoming platforms. In fact, one really easy way to start looking for software that might solve a specific problem you have is to go to the CM software you use and review their partners. Procore specifically has an entire marketplace with hundreds of partners – in this case, not only do these companies work with an API, they will have been vetted by Procore, so you have some degree of confidence the software will work as advertised.

Many other companies, though, aren't formal partners but have an API that'll work with the CM software. This is one of those instances where a pilot is a good idea, because having an API connection doesn't mean they've quite figured out how to work well with the CM software you use. As an example, Procore has over a dozen API "categories" just for daily reports. The way Procore chose to organize daily report information might not be the way your potential vendor did, so you'll want to see if they can make it work. Usually they can, but it can cause real delays if they have to figure it out during your deployment.

The difficulty here is that many specialty contractors won't have only one CM software they work with, as they'll use what different GCs require for different jobs. We'll discuss that as follows.

## Do You Have the Team to Manage It?

We are just now on the cusp of some really interesting software that can do amazing things, like reality capture, artificial intelligence, and more. Very often, these amazing things will require a semi-dedicated resource to manage.

For some technologies, like drones, this is obvious. But for many types of software that just digitize a paper process, or gather some kind of information, it is not obvious at first that someone, usually a small group, will need to shepherd the software for a year or more so that your teams get used to it, and start to *really* change their processes to take advantage of what the software can do.

Nothing magically works out of the box, and much of the value you'll get from new software will come from unexpected changes in how people work, what questions you can answer, and what decisions get made more quickly or confidently. That takes time and focused effort, time that should be allocated up-front and consciously, or you risk failure of the pilot and deployment that follows. In today's world, where technology has become part of competitive advantage, not getting the most out of your software and technology choices puts you at a disadvantage to other firms.

## Specialty Trades Versus General Contractors' Use of Field Software

This final section relates to the difficulties faced by specialty trades in making almost any technology decision. For a start, trade contracting companies are often smaller than GCs, so will rarely have big innovation teams.

But the bigger issue is that specialty trade contractors are hired by a GC, who runs the site, and determines which technology is used. As a result, trade contractors usually need to support multiple CM software products at once, and are very often using technology that has been developed for GCs, not for them. This is changing as Procore, Autodesk, and others have been increasingly focusing on the trades, though there is work to be done to capture the higher level of jobsite detail that a trades crew will need.

There are some software vendors, like eSub, who very specifically target specialty trades. In the case of eSub, they provide most of the same functionality that a CM provides a GC, just geared towards the unique issues that specialty contractors face.

Regardless of what software you are looking to adopt, the discussion above is just as valid for a specialty contractor as it is for a general contractor, because most of the watchouts and steps I've outlined are more about the difficulties facing any company adopting software than they are about the specifics of how it is used.

In general, you cannot assume that the developers of software understand your work process, skills, and conditions under which you work. To expect that a small software vendor will understand every trade is unrealistic, but you can work with them to take some of these innovations and make them work for you – just like almost everyone has used Excel for years, despite the fact that there is no "Pipefitter's Excel."

A good approach is to ask that they meet you where you are, from a requirements and comfort level standpoint. Go through Steps 1–4 mentioned earlier, but focus also on your specific trade. Ask for cases and references for yours, or adjacent trades.

In all field software, just like in all construction technology, what is most important to remember is that all of these technologies and products exist *for you*. Skilled workers bring the experience, intuition, and judgment, which is what makes the built world stand. Tools, no matter how many flashing lights they have, just help you do that a little better.

More has been written about BIM than any other technology in AEC. Most of what's written relates more to architecture than construction, because most use of BIM was from the design side of things. That is changing, so let's view BIM from the construction perspective.

## BIM

BIM is yet another word that gets used without a clear definition, so let's start there.

BIM, of course, means "building information model." That means a collection of the information about a building. It does not mean a Revit file, or a Rhino file, though both can produce a BIM model.

BIM is often mistaken for the geometric, 3D models that modern BIM almost always includes. But BIM isn't the 3D model, it is meant to be exactly what the name says – a model of the information about a building.

As Dana Smith and Michael Tardif point out in their fantastic book *Building Information Modeling*: "The geometry of a building represents only a small percentage of the total body of useful information about that building. A genuinely comprehensive building information model would encompass not only geometry but all of the information about a building that is created throughout its useful life."

This definition is troublesome, because "all the information that is created throughout its useful life" means everything from design through construction to handoff to facilities management through to demolition. There is no agreed-upon format for that much information, no standard for how everyone who needs their kind of information would be able to ignore all the information they don't need.

For a construction crew, the "useful life" means since the most recent approved plans; the history of design changes is not something the field crew should even see, as it's already hard enough to keep everyone current on the latest plans.

What matters to construction about BIM is that it has begun to formalize the design capabilities that specialty trades have always had for their own assemblies and work to be put in place. In fact, more important than BIM per se is the movement towards virtual design and construction (VDC) in the trades and general contractors.

As more and more contractors learn to work in Revit, the ability to conceptualize proposed solutions and communicate those proposed solutions to others is hugely advanced, over earlier paper-based workflows. It is just faster to see what people mean when you can see it in 3D.

Three big developments in BIM that are worth understanding:

1. BIM for clash detection is essentially automatic – VDC workers still need to do their jobs, but more and more software solutions for this once painful exercise have made it less so.
2. CM software includes BIM that can be viewed in the field. Procore, Plangrid/Autodesk, and others now offer easy ways to view the latest BIM model on your mobile devices, wherever you are. This adds a layer of intuitive understanding of the work to be put in place that is simply better than 2D plans. What is also helpful, though, is that these programs do not make you choose – you can view both 2D PDFs and 3D BIM models, depending on which you find more intuitive, given your experience level with BIM.
3. Computer vision and augmented reality have made it possible to orient the model against an actual jobsite, so you can view the model superimposed over the actual work put in place, both to see progress and plan for what's next. Trimble's collaboration with the Microsoft Hololens is a great example of this, as is Holobuilder, Openspace, and some other startups. The ability to make digital plans available in this highly intuitive user interface holds great potential for continued development of tools that augment the intuition of the trades and GCs on site.

BIM as a coordinating tool is only now beginning to fulfill its promise, and as startups and major software platforms alike make it more available onsite, more linked to the actual geography and work put in place, the potential for time savings and efficiency overall is huge.

# CHAPTER 5

# Industrialized Construction

There is a tension in the construction industry, between the need for improvement and the need to keep risks low. In any inherently risky industry, like construction, the reflex is always to keep what works and build on it, not change it for less understood, less intuitive methods.

Software will digitize the current workflows, and that will make a difference in terms of safety and efficiency of construction jobs. However, to make a real leap in safety, efficiency, and productivity, and attract a new generation of workers who have other options and would prefer to avoid spending the entire day outside, industrialized construction is going to have to become more prevalent. The 2020 pandemic can be expected to add to this trend, as it is much easier to enforce social distancing in a factory than it is on a jobsite.

And across different sectors, we are beginning to see this.

## Types of Industrialized Construction

What do we mean by "industrialized construction?" Here one term is used to mean lots of things, so let's start by defining what industrialized construction means.

Industrialized construction is a collection of methods that apply manufacturing techniques to construction, and includes the following:

1. **Prefab**: One-off assemblies done at a location other than where it is finally to be put in place. Prefab can be done offsite, or in a staging area on the jobsite. Generally, prefab is one trade doing assemblies for one job, and the point of doing it separately from the final location is the ability to decouple the schedule of building a given assembly with the on-site schedule, which is important for specialty contractors.
2. **Flat pack**: Assemblies that are created in bulk, where the same assembly is created multiple times and packed up for transport, then unpacked and installed on site. Flat pack production benefits more from the economies of scale that give manufacturing its cost and productivity advantages, though the quantities are nowhere near what most product manufacturing operations produce, so the benefits, while real, are not as substantial.
3. **Modular**: A multi-trade, fully functioning part of a building that is produced in an offsite factory. Often, this is a bathroom, hotel room, or other building section where variation in size and features can be limited without sacrificing a building owner's goals.
4. **Volumetric**: Modular construction that is fully connected to the building's design, systems, and operations. This will mean that the building was designed to use the modular design on every level.

These levels become more efficient, digitally driven, and automatable as they progress from prefab to volumetric, and the assumed quantities go from one-offs to hundreds of thousands of similar increases along the same axis.

IC will not throw you out of a job – at its simplest, IC is just doing your job offsite, but even the more complex versions of IC require an understanding of the trades to work. According to Don Metcalf, prefab manager at Nemmer Electric, the benefits he sees on real jobs include:

- **Time savings**: You can often get things done more quickly in a controlled environment like the prefab shop than on a jobsite.
- **Flexibility**: Not being tied to the schedule of the jobsite means you can do prefab work when the weather is bad, and schedule it when convenient for other reasons.
- **Cost**: Because it is faster, it is cheaper to produce. There is also the opportunity to bulk purchase commonly used materials, like conduit or sheet metal that the company will need for multiple jobs, further lowering cost.
- **Quality**: Doing the same thing in the same place means you can get better at doing it. It is also much easier to inspect work done offsite.

Prefab and industrialized construction can also threaten the pride in a job done well, which is part of being a construction professional, because it doesn't feel quite the same to build it and then install it as it does to figure it out onsite. In time, that sense of pride will include other things, like designing better sub-assemblies, working out better ways to fabricate, and other kinds of problems that are what pride in a job done well are all about. Managers would do well to keep in mind that money isn't the only reason folks go to work every day, and recognize that the desire to *create* something is important.

Understanding these four levels of IC leads to the next question – how does this fit with the design and building process that includes submittals, RFIs, and change orders? How does that process work with a design process that is more and more like industrial engineering, where *how* you make something is as much a part of the design as *what* you make?

Modern construction has, in effect, two sets of designers: architects who are recognized as the design professionals, then the skilled trades who design sub-assemblies formally, and informally design the specific work that is to be put in place. In effect, trades are like the industrial manufacturing engineers who have to figure out how to make the designs they receive into something that can be built.

There's a phrase from manufacturing that will resonate with anyone who's been in the field, and had to try building something that just didn't work: "over the wall," which refers to the oldest and least efficient way to design and produce a product. This sort of thinking led to a sort of revolution that borrows from the lean production movement: design for manufacturing and assembly, or DFMA.

## DFMA

DFMA is really about two things – designing products so that they can be produced, and then designing them so these products can be assembled efficiently on the jobsite. Many readers will have come across statistics about how much of a job's cost is driven by decisions at different stages – with upwards of 70% of a job's cost decided in the design phase. That is when things like roofing, square feet, number of floors, and overall shape of the building are figured out.

The same is true when we're designing any of the levels of IC, but especially modular and volumetric, where the quantities will be large enough that really thinking about how to manufacture the parts or assemblies will have a huge impact on cost, efficiency, and quality. Just like on the jobsite, we need to worry about what tools will be needed to manufacture an assembly; in fact, the entire design process more resembles that of a car factory than of construction.

There is software that can help with this process; in fact, the term "DFMA" is trademarked by Boothryd Dewhurst, and they offer software that is explicitly aimed at taking cost out of both the manufacturing and assembly processes.

### Design for Manufacturing

Because you are creating something that will be produced over and over, it is worth thinking about each part of the process, each element of what's being created, and asking where we can find efficiencies. A partial list of where efficiencies can be found includes:

1. **Materials** – do we need a given grade of metal, thickness of pipe, and so on? Will certain materials be easier to fabricate at scale than others? Do we have options in a fabrication shop that we would not in the field, because we have better/bigger/more powerful machines?
2. **Cycle time** – can we design for shorter cycle time?
3. **Special tooling** – how much of the process can use off-the-shelf tools?
4. **Direct labor** – how much time is spent touching the workpiece?
5. **Indirect labor** – how much time is spent supporting the production process?
6. **Waste** – how much material gets wasted?
7. **Perishables** – drill bits, grinding wheels, and so on.

By carefully considering each of these elements and making tradeoffs as necessary, value can be maximized for both the building owner and the contractor. For this to happen, though, the process of engineering needs to involve all stakeholders, not just the specialty contractor doing the work. DFMA provides the frameworks to really examine what costs a workpiece might create. The next framework expands on these elements to include other stakeholders: DFA.

## Design for Assembly

Whatever we fabricate or manufacture, it will need to be transported to the site, then to the place where it will become part of the building. That means getting on a truck, getting off a truck and across a jobsite, and then up and into a floor of a building. Sometimes the entire job schedule needs to change to accommodate the installation of a large modular piece, for example a mechanical room. More often, the modular component or prefabbed assembly must accommodate the path it will take to get to the jobsite.

Considerations will include:

1. **Skill of onsite labor:** Not everyone knows how to install prefabbed units. It should be designed to match the expected skillset, or be easy for them to learn quickly.

2. **Number of onsite workers**: Larger assemblies might mean more team members.
3. **Tools needed to assemble or install**: Is there a special toolkit needed? Or can it be done with standard tools?

## Other Considerations of DFMA

The general point of DFMA is to design work to lower the overall cost and complexity of producing, transporting, installing, and owning work that is to be put in place. That should include how easy the work is to maintain, how much we've reduced the environmental impact, and how easy it is to check for quality and wear and tear over time.

DFMA is an emerging skill that is reaching maturity in construction, though the sheer range of building and work types in construction means this will be an area for development in the coming years.

## IC Tools and Processes

Producing things in a manufacturing facility means you have options for production processes that are simply not possible on a jobsite. Here are a few that are worth understanding:

1. CNC (Computer-Numeric-Control) is machine-driven cutting, drilling, and fabrication of workpieces. This is often done on simple shapes like flat sheets, rods, and pipes.
2. 3D printing uses plastic or metal that is deposited in layers to create unique shapes. Early quality issues have given way to machines that are so good that orbital rockets are made with 3D printed engines. 3D printing has been used for specialty parts more than other uses, but in time it will find more and more applications.
3. Robotics have been used in manufacturing for many years, but are less of a feature in construction-based IC processes because the expense of most robots means they need to have a large production volume, which most construction IC projects currently don't satisfy.

4. Assembly lines are the most basic part of manufacturing, and are a new way to think about construction work, as each person does just part of the assembly, then passes it to another. This allows for hyper-specialization and efficiency, but is not what most trades signed up for and there can be resistance.

## IC Software

Not surprisingly, some of the software that comes from manufacturing can be applied to construction but, just like assembly lines, it is really built for much higher volumes and larger datasets that enable statistical control that's just not appropriate for a construction job or set of jobs.

There are two broad classes of software that can manage industrialized construction tasks: design software and management software.

On the design side, Revit, the primary BIM software used for most VDC, was not created for fabrication work, and many contend it is not really made for that use. However, others argue that is because too many people haven't learned to use that function of the tool, but that it is more than sufficient to handle fabrication.

For CNC and control of machines, though, you're going to want to explore specific design software made for that, like Fusion360, Solidworks, Rhinoceros, Creo Parametric, and AutoCAD Mechanical. While each of these has different features, all are made to create 3D models that can then be sent to fabrication machines, whether they be CNC, 3D printing, or others.

On the management side, there are newer entrants that are purpose built for managing different levels of the IC process. Three options are:

- Manufacton – much more of a process management solution, this software allows for collaboration, scheduling, and communication for IC projects.
- KitConnect – created by Project Frog, this software connects design to manufacturing to delivery. Intended for use within a BIM environment, KitConnect is a mature product.
- FactoryOS – as the name would imply, FactoryOS is a management tool for fully fledged factories, with a special focus on homebuilding.

Industrialized construction has made great strides in recent years, and as we emerge from the Covid-19 lockdowns, it is expected that factory settings will allow for recruitment of the elusive millennial worker, and allow for the productivity that building the world of the 2020s and beyond will require.

# CHAPTER 6

# Machine Learning and Artificial Intelligence

"We propose that a 2 month, 10 man study of artificial intelligence be carried out during the summer of 1956 at Dartmouth College in Hanover, New Hampshire. The study is to proceed on the basis of the conjecture that every aspect of learning or any other feature of intelligence can in principle be so precisely described that a machine can be made to simulate it. An attempt will be made to find how to make machines use language, form abstractions and concepts, solve kinds of problems now reserved for humans, and improve themselves. We think that a significant advance can be made in one or more of these problems if a carefully selected group of scientists work on it together for a summer."

—"A Proposal for the Dartmouth Summer Research Project on Artificial Intelligence" (1955): J. McCarthy, Dartmouth College; M.L. Minsky, Harvard University; N. Rochester, IBM Corporation; C.E. Shannon, Bell Telephone Laboratories

A s the quote above shows, artificial intelligence (AI) is not a new idea; in fact, we've been trying to make it work for almost 70 years. And more than once, like now, people have thought AI was about to become human-like in its capability. The pattern is the same each time – we have some early wins that are really impressive, and researchers and the media all think this early progress will continue, until we've created a true AI.

And of course, so far no one has come close to creating an AI that can think, in part because no one really knows how hard many of the things we call "intelligence" really are – in fact, unlike almost any other kind of technology, we don't understand the underlying mechanism. The basic idea of "intelligence," and the specifics of how thinking happens, aren't understood in humans, much less in a machine we create.

In the quote at the start of this chapter, I include all of the authors of that first paper because two of them are about as good as it gets when it comes to computer science: Marvin Minsky founded the MIT media lab, and Claude Shannon laid the foundations for all modern communications. Very smart people have been underestimating how hard this is from the beginning, so it's no surprise that it keeps happening.

## What Is Machine Learning and Artificial Intelligence?

Machine learning is not the same as artificial intelligence, and there have been attempts at making AI that did not use machine learning. In fact, there have been three waves of AI: the first two followed by disillusionment and what is often called an "AI winter."

However, almost everything you'll hear referred to as "AI" today has machine learning at its heart, so let's make some clear definitions:

- **Machine learning:** The study and creation of software-based algorithms that build a mathematical model based on data, that can make decisions, predictions or perform tasks without being specifically programmed to do these tasks. It contrasts with ordinary software in a few key ways:
  1. Machine learning aims to create a system that learns from experience, getting better as you use it
  2. Machine learning trains its models on huge quantities of data, often millions of data points
  3. Machine learning can handle dramatically more complex tasks
- **Artificial intelligence:** Software that mimics some aspects of human mental capability, including machine vision, natural

language understanding, decision making, and pattern recognition. AI can be built using machine learning, though there are other methods that have been used to try to mimic human capabilities, like "expert systems" in the 1980s.

A simple way to understand the two is that machine learning is a set of techniques for building AI, and AI is the set of applications and products that are available for users.

## Why AI?

In the past decade, there has been a lot of hype about AI. We consistently think AI is a bigger opportunity and a bigger threat than it is likely to be, and these hype cycles inevitably get to a place where speakers and writers think they are being forward looking when they make provocative claims about dangers and opportunities. Let's take a realistic look at the opportunities.

Modern AI can do things that no other software can come close to doing. In fact, compared to other software approaches, it can seem almost magical. AI can predict what will happen next, it can recognize faces from amongst millions. AI can find patterns in oceans of data, it can protect your bank account, and can weed out spam emails. AI can recognize language and generate seemingly intelligent responses.

When it comes to interfacing with a computer, AI has completely revolutionized the possibilities, and in some cases the realities, of user experience. For example, we now have voice interfaces, gesture interfaces, and software that can guess what you might want next.

For larger systems, like construction management software, you will be able to pull together and analyze your data so that bigger pictures can be seen. AI holds the promise of empowering managers to really understand what is happening now, and what is likely to happen next, within their businesses.

For smaller solutions, we are seeing everything from voice operation on the jobsite to machine vision identifying social distancing practices, to systems that can knit hundreds of pictures into a real-time, as-built record of the jobsite. These specific applications of AI to the problems of construction are going to keep coming, as the technology matures and more people are able to create new products without needing to have a PhD.

AI is a self-sharpening tool – that rare human invention that can improve over time, can actually *learn* from being used in the field. That is so unique, so powerful, that it has very much caused otherwise rational minds to expect amazing improvements to continue, which they have not.

Perhaps most importantly, because AI is software, once you have trained a model to do something, say check BIM models for clashes or constructibility, you can make copies of that model for free, meaning you can immediately and costlessly create an army of BIM-checkers.

If humans had been employed to be that BIM-checker, you now have a huge number of human salaries you do not have to pay. And an army of BIM-checkers no longer checking BIM.

At least, that is the idea.

## AI's Weak Spots

The practical reality is that AI is never that good, never that reliable, and never that trustworthy on its own. Human beings with decades of experience make mistakes, but we know *why* they've made a mistake so we can predict and correct for likely mistakes. Because AI is so different from a human mind, we can't always predict what mistakes it will make. This doesn't mean we can't use AI, it just means the hype is not justified, and that real applications take much longer to get right than non-developers think.

There are two reasons AI can fail that I want you to understand: Edge cases and bias.

### Edge Cases

Recall we talked about "edge cases" in Chapter 2 – those instances where a user, or just the data, don't quite fit what you designed the software to handle. That happens with AI, because we've fed the model a big set of data, but that data will not include every case the AI might encounter, at least not at first. So when we're testing the AI we realize it cannot handle certain edge cases, and we need to go get more data.

Edge cases are an issue with all software, but they can be more of a problem with AI, because we've become so used to AI performing what looks like human-level feats, we assume that AI has the same breadth of thinking and cognitive capability as a human, and it does not. And we can assume that early, amazing demonstrations will continue to improve at the same pace as the early stages, and that never happens.

Autonomous driving is a great example of this: we had some early wins that showed great promise, and lots of breathless predictions about how the whole world would go driverless by 2019. Lots of concern over truck drivers losing their jobs, and so on. Except the world is full of edge cases, and the models still cannot handle real driving, especially on residential roads where the stakes are so high, with children and families in much more potential danger than a highway.

Edge cases and their ability to defeat even the most sophisticated AI models illustrate an important idea in AI: how accurate do you need the AI to be? For a book recommendation, we can accept accuracy of 80% or so. But for a car driving in residential neighborhoods, we need accuracy of 99.9% or higher, and that is *very* hard.

## Biases

Machine learning and the AI products it can create depend on lots and lots of data. But what if that data doesn't reflect reality? What if you only have men in your data set? What if you don't have enough of a given ethnic group? Or what if you include data that only shows 50-year-old men late to the jobsite, with under 50s all showing up on time, just because that's what you have in your daily reports? Your AI will learn to view the world in a biased way: it will not recognize, or mis-recognize the women, or the minority that is under-represented. And it will predict lateness in over 50s automatically, without asking why.

Correcting biases is one of the biggest reasons humans are going to be needed to manage and oversee AI for a long time. AI doesn't "think," it just correlates data. So we need to apply judgment to what AI does to make sure it doesn't go wrong.

Artificial intelligence as applied today does some amazing things, but it is not "intelligent" in any meaningful way, and still just software. To understand what all of that means, let's talk about how machine learning works its magic.

## How Machine Learning Works

Machine learning uses data to "teach" software to recognize things in the world, based on a huge number of examples. The fact that we can create software that can learn is a bigger deal than it might seem at first, and speaks to perhaps the most important takeaway for really understanding AI and machine learning:

> **Instead of trying to compare machine learning and AI to human intelligence, instead compare it to what non-AI software can, and cannot do.**
>
> **Machine learning is software, not intelligence.**

There are three main machine learning approaches, each of which deal with data differently and are used for different types of problems. These are:

1. Supervised learning
2. Unsupervised learning
3. Reinforcement learning.

Understanding how these work helps you understand how machine learning works in the real world, and why data is so important to machine learning and AI products.

### Supervised Learning

Machine learning uses tons of data to train its models. So how do we shove a huge pile of data into some software and hope it learns something?

The easiest way is to have that data labeled, or "supervised." Most of the data that is in a real, production model of AI out in the world uses supervised learning, because it is a much, much faster way to train a model.

As an example, if I'm training a model to recognize five dog breeds, I will prepare as many as a million photos of dogs, with each photo labeled as "Husky," "Labrador Retriever," "Bloodhound," "Bulldog," or "Poodle." Each time I feed an example through the model, it will learn a little more about what makes a bulldog different from a poodle, because it can refer to those labels.

We call this "supervised" not because a human actually supervises the training, but because the model has exact reference labels for every datapoint, and because someone somewhere thought it sounded better in an academic paper to call it "supervised" instead of just "labeled."

A million examples sounds like a lot, but think for a moment about how *much* we are teaching some of the models, how subtle the differences AI products out there can handle. For example, Facebook, Apple, and of course Google have products that can recognize you, personally, from a photo or video feed. Of the millions of people that have your height, skin color, hair color, and so on, AI can recognize you.

Compare that to someone trying to write rules that could recognize you, instead of using a mountain of data to train a model like we do in machine learning. It may be technically possible to write software that could do this, but it would be insanely hard and take an incredible amount of time. In fact, in the 1970s researchers were trying to do something like this, using what were called "expert systems" at the time. Incidentally, this was part of the second "AI wave."

These researchers were trying to create a system that would recognize when a high school math student was doing well, or poorly, at math. The system was to recognize from their answer and other cues what they needed to hear next, what refresher or other studying would be best. The project was abandoned when the researchers realized they had to program over 500 hours per minute of student time, or 1,250 days for one hour.

Another example is voice technology. Some readers might have used Dragon voice to text in the past, and if so, you'll recall how inaccurate it was, and how it needed to hear your specific voice for a training session before it would start to work. There is a reason we barely heard about voice until 2011 or so. These systems were actually very clever, but they didn't learn from data like modern, machine learning approaches do.

In contrast, Siri, Alexa, and other AI voice products do learn, constantly, from hundreds of millions of users, and as a result they get better all the time. Whatever frustrations you have with these systems, it will not be because they don't understand your English, it will be because there is no intelligence behind them to understand your meaning.

This learning by connecting examples to labels is a lot like regression. You may recall from statistics class the simple idea that if you have a bunch of data points, and you want to understand if one is related to the other, you plot these points on a graph, where the *x*-axis is one factor, and the *y*-axis is another.

**The Construction Technology Handbook**
*Submission by Hugh Seaton*

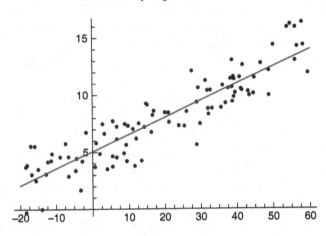

For example, if you wanted to know if the air temperature has an impact on humidity, you might take a few dozen readings outside, where you look at the humidity percentage and temperature at each of these readings.

You wind up with a few dozen dots that kind of show an upwards trend, suggesting that higher temperatures lead to higher humidity. That's great, but what you *really* want to know is, for a given temperature, what's the likelihood of high humidity?

Regression is a process where you can figure out what the relationship is between these two factors, expressed as an equation.

This is just like supervised learning, where one axis is the label, and the other axis is the data point – except instead of two axes, you have a huge number of factors.

Regression is good as an analogy so you can understand labeling, but the reality of modern machine learning is that the math is quite complex, and thankfully very few people actually building AI products need to really understand it at the deeper levels.

In practice, especially for images and now language processing, you'll use an existing model that has already been trained on millions of examples. These models are general enough that engineers are then able to retrain it on a specific, smaller data set, and your developers can access them through big companies, or get open-sourced models from GitHub.

For example, you can take an image recognition model from Microsoft, Amazon, or Google. Then you can import a few thousand images, let's say of a kind of tool, like a wrench. At first, the model will have relatively low accuracy, but as you give it more and more examples, it will get more and more accurate. To be clear, the model doesn't run itself, someone needs to be tweaking some of the settings, but training a model for use like this is much more feasible than finding a million pictures of wrenches.

Three points to remember about supervised machine learning:

1. Supervised learning means the data is labeled
2. Models require many millions of examples to train from scratch
3. Modern models, especially for images and language, can sometimes be trained from existing AI models

What that means for you, the construction professional, is that the value to your organization, and ultimately to you, of consistent, well prepared data is potentially huge. You will be able to build your own models to predict the likelihood of safety issues, of rework, of RFIs, and many other critical issues in the future, but the system that can do that will need your data to train.

## Unsupervised Learning

Sometimes you get a huge quantity of data and you don't even know how to label it, because you don't know what bucket to put it in. An example of this is when you don't know what causes mechanical failures, but you have sensor data from different parts of the machine and system. You need to feed this information, usually many millions of data points, to start understanding what sensor data correlates with other sensor data, so you can begin to create labels that, in turn, will allow you to design a system that can predict these failures when they are becoming likely.

That is what unsupervised learning does. The software will take a huge quantity of data, and will figure out how to make classes, or buckets, for that data. This sort of thing is actually rarely used for production AI systems, but is instead the way that very large models are designed. For example, in 2018 Google was working on a new way to train a model for understanding language. Those models are called "Natural Language Processing," or NLP. Being Google, they have more data than anybody, and have more processing power than anybody.

Google used unsupervised learning to take this enormous dataset, crunch it into buckets, and train its BERT model to recognize and make use of complex language sentences in over 70 languages. Google's data scientists had figured out the approach they wanted to take, they had an architecture. But they didn't know the "buckets," they didn't know how to label such an enormous amount of information, so they let the system do the bucketing for them.

It is estimated that Google spent about $20 million just training this model, using their own servers, showing just how big an endeavor unsupervised learning can be.

## An Aside on AI Models

In the previous example, Google's team had an "approach," otherwise known in AI as a model. Models are an important concept, one that gets missed in popular discussions of machine learning and AI. The model that gets chosen, the algorithm it is a part of, is most of where the magic really comes from. As we've seen, huge quantities of data are needed to make machine learning work, especially for more demanding tasks like machine vision, natural language understanding, and others.

In the case of BERT, Google was improving on a long line of model types, with fun names like Word2Vec, ELMo, and ULMFit, and was improved upon later by OpenAI's GPT-3. Each of these is like a version of a car engine – they work basically the same, but parts get upgraded as engineers learn and try new things.

We can only train an AI to do things its model allows it to, and that is crucial to our understanding of AI overall, and to recognizing the limits of any AI implementation or product you're going to deal with.

Machine learning, and therefore AI, cannot do anything it has not been designed to do. It cannot learn entirely new skills that it wasn't designed for, it cannot invent new capabilities. The popular press frequently misses this point, as they do with the bigger point of how hard it is to make a real model actually work. Training a model the size of BERT, for example, will have taken hundreds of man hours for data scientists to tune the model as it was being trained and re-trained. It is hard, grueling work as you are basically building, testing, failing, then rebuilding to repeat the cycle until it works to your required level of accuracy.

## Reinforcement Learning

While supervised learning is the machine learning workhorse that's being used in most production AI products, and unsupervised learning is how we've been exploring much bigger, harder, and unstructured data problems, perhaps the most exciting machine learning approach in recent years has been reinforcement learning.

Supervised learning and unsupervised learning are good at dealing with one-time processes. There is no memory of what has happened in the past, there is no way you can have a sequence of steps, for example. Supervised and unsupervised learning are perfect for one-off tasks like recognizing a photo, analyzing a 3D scan, understanding a paragraph, and so on. Each of these happens once, then the system resets and is wiped clean. If I show an image recognizer images of, say, a gravel pile that's 5 feet high, then an image of that same gravel pile at 7 feet high, then again at 10 feet high, the image recognition algorithm will not do anything different each of those times it recognized a gravel pile. The fact that they are happening in sequence is not information that the model has a place for – it is irrelevant.

Many things we want software to do require a series of steps. Someone does something, it causes a change in the world, then they do another thing, or someone else does something in response to the first person's actions, and so on.

Reinforcement learning is called that because the AI system is given a reward every time it gets a step in a process correct, so we can train individual steps in a series. The algorithm doesn't get "rewarded," in any sense we'd recognize, it is just a series of mathematical steps. We run the algorithm hundreds and thousands of times, each time rewarding each of its substeps a little if the algorithm does a better job, and less if it doesn't improve.

As computers have grown in power and our ability to simulate different processes has too, we're able to use reinforcement learning by simulating a process, and training hundreds of thousands of versions of an algorithm until we find a version that works best.

The power of this approach became clear in March 2016 when Google's AlphaGo beat Lee Sedol, the 18-time world champion of the strategy game "Go." Invented in China over 2,500 years ago, Go is a chess-like board game that is played across East Asia, and is considered the hardest such game there is.

Reinforcement learning makes sense here, because to win you need to correctly execute a series of moves, in response to what your opponent is doing. The large number of possible moves and the incredible possible complexity of Go make it a particular challenge.

A Go board is made up of a 19 by 19 grid of squares, and each player has either black or white pieces that are all equivalent in rank, unlike in western chess. The goal of the game is to capture territory by encircling and removing the other player's pieces. The rules are simple and relate to how many moves, turns to take, and so on. Because the rules are simple and the pieces are all the same, the game is almost infinitely more complex than western chess – for illustration, it is estimated that there are $10 \times 10^{170}$ different ways the board could be arranged – an impossibly large number that is actually bigger than the number of atoms in the known universe. Go is a tough game to master.

That huge number of possibilities means that you cannot just do a computer search for all the possible next moves and their follow on; you need to be able to understand the game at a deeper level. When IBM's Deep Blue beat the world chess champion, Gary Kasparov, in 1997, it did not win at chess, it won at searching for possible moves at lightning speed and finding the best match. That approach doesn't work for Go, which is why most people thought it would be decades before AI could beat a human, but beat him it did.

Reinforcement learning is not easy to get right, but it points the way towards AI helping everything from scheduling to automation of certain tasks by software. It does not mean that what humans do can be automated quite that easily. We'll return to that below and in the next chapter.

## Deep Learning

AI has gone through three waves. The first wave used a collection of software approaches to approach AI as a logic problem – often referred to as GOFAI, or "Good Old Fashioned AI." We still reference many of the ideas and breakthroughs from this period, as they helped to frame the problem, even proposing some of the ideas that later proved important. However, the math, computer science, and most importantly, actual computer hardware and software of the time was nowhere near good enough to take these ideas forward, so the field fizzled out in the 1970s.

In the late 1970s and 80s another approach was tried – instead of operating at a higher level of logic, researchers tried to set rules. Known as "expert systems," these approaches showed early progress until they were confronted with the complexity of real applications. Expert systems failed because you just cannot write enough rules for software to deal with the real world. The end of the 1980s saw the excitement for AI falter again.

But all along, there have been computer scientists working away at different approaches, and one of those, deep learning, turned out to be important.

## Deep Learning Defined

Deep learning is loosely based on the way the human brain works, so let's briefly understand what that means. The human brain is made up of about 85 billion neurons. Each of these has many inputs, called dendrites, that connect to lots of other neuron's outputs called axons. Each neuron only has one axon, that in turn connects to many other neurons' dendrites. When enough of the dendrites of a given neuron receive enough signals from their upstream neuron partners, that neuron will "fire" a signal, which will then go down the length of the axon, and in turn send a signal to many other neurons.

The typical neuron is connected to an average of 5,000 other neurons, and as many as 50,000. So the brain is this huge network of neurons firing at each other.

The important thing to understand is this idea of many neurons sending signals into one neuron, and the sum of those signals needing to pass a threshold to cause the neuron to fire. That threshold is reached in two ways.

Either many, many neurons send signals into the neuron, and they pass the threshold. Or fewer neurons send signals to the neuron, but each of these sends a stronger signal, again passing the threshold.

Learning happens because neurons that get fired together often will have their connections strengthened, so they can send a stronger signal.

These neurons sending signals to another neuron creates a hierarchy, and hierarchy continues throughout the brain. In other words, we have layers of neurons that go from the "bottom" of whatever we're seeing, hearing, or thinking about, up to higher levels of perception or thinking.

Let's illustrate that with an example of how your brain would see a box – the neuron-based steps that allow your eyes to send data into the brain that gets processed so you see that box. It sounds simple, but actually is very sophisticated.

When your eyes look at a black box like this:

Some neurons that respond to black will pass forward information about whether the eye is seeing black at a given point to another set of neurons. Those next neurons will take the black spots and blank "spots" and add them up, so some neurons will process an edge, some will process a long length at 90 degrees, another at zero degrees, and so forth. It turns out that in the brain there are groups of neurons that are good at recognizing basic shapes, edges, and so on. These groups are trained to see shapes and relationships, so they fire when they see them.

Next, these neurons pass their signals forward to neurons that recognize angles where two lengths connect, then again these will pass forward their signals to neurons that recognize squares, and so on until a box is recognized.

Each of these layers has been trained over the years to recognize things it sees in the world, by seeing them over and over.

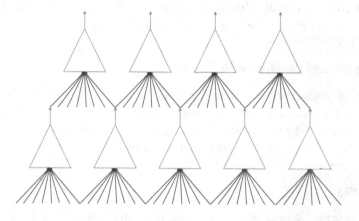

That is what deep learning does.

In deep learning, a number of "nodes" are created in software that work like a human neuron. They take inputs from something like a camera, and pass them into a number of these nodes. Based on the strengths each of these nodes has with the input source, some of them fire and pass on information to the next layer, and some do not. This happens a few times, often not more than 15 layers, then an output is created. Usually that output is classifying something, whether it's a picture of a dog, a sentence, or a pattern in financial data.

The power of deep learning is that, each time you give it an input that has a label, it can "learn" because an incorrect classification will cause the whole network to reset the strengths between these nodes.

So say, for example, you have three layers in the network, and we pick three neurons from layer 2 going into a single neuron on the third layer. During training, a picture of a dog is processed, but the network thinks it is a human. That gets processed as an error, and the connections between neurons 1, 2, and 3 will be changed. What is helpful about the way this is done in practice is that the training will know in which direction the strengths need to be changed – stronger or weaker. How much they are changed is set by the data scientist doing the training.

This process happens over and over again until the model is good enough at the task you are training it for – often training can be hundreds or thousands of these cycles.

Because we have split the problem of identifying something into a hierarchy like this, we are able to be incredibly flexible, and also incredibly powerful. That is the magic of deep learning, and why it is behind much of what works in this third wave of AI.

## Imagenet and the Birth of Modern AI

What is remarkable is that the deep learning approach has been known for decades, but we weren't quite able to make it work like it does now. Three changes have made deep learning possible: more data, better computers, and better models.

## Big Data

The internet, World Wide Web, and telecommunications have utterly changed how humans consume, produce, and share information. Just think about images – what used to take a special camera, special film, and at least an overnight process to develop pictures now happens effortlessly, instantly from your phone. Similarly, where it used to require expensive copying and physical mail to share an image, now we send pictures around through email, messaging, social media, and so many other ways we take it for granted.

This effortless creation, copying, and sending is true of almost every kind of information. As a result, the worldwide creation of data is so big that once again it involves numbers we cannot really understand as humans. Personally, I love the names that are used – according to the World Economic Forum, the human race has created about 44 zettabytes of data by 2020. How do we get to a zettabyte?

If you have a million bytes to make a megabyte (MB), 1,000 times that is a gigabyte (GB), and 1,000 times that is a terabyte (TB) – all numbers you've probably worked with for a file, computer RAM, and hard drive storage, respectively. A thousand terabytes is a petabyte, which is the size of data Facebook generates in a day (about 4 PB in 2019).

A thousand petabytes is an exabyte, the sort of quantity the world creates in one day, estimated to be up to 265 exabytes per day in 2025. A thousand exabytes gets you to the zettabyte, of which, you will recall, the world has something like 44. But at this rate we'll blow past 1,000 zettabytes, where we'll get to the yottabyte.

Personally, I would like an invite to the next meeting where data quantities are named – it looks like fun.

What is at least as important as the amount of data is our ability to manage all of this data. We've innovated new ways to store, search, and process data that makes it even more useful, often years after it was collected, for the training of AI models.

## Computing Power

These unimaginable piles of data take effort to process. The rise in computing power over the past 70 years is well established and surrounds us. Many of you will have heard of "Moore's law," which says that the processing power of computer chips will double about every 2 years. This grows out of an observation by Gordon Moore, one of the founders of Intel, in 1970, that the size of the transistors that make up a microchip will shrink by half and the cost will also be reduced by half, in that 2-year period. The practical result of that is the doubling of power of microchips every 2 years.

That sounds cool, and doubling is impressive. The real power of constant doubling is that it compounds. Perhaps, for example, you've heard the thought experiment about doubling a piece of paper. The question at the heart of the experiment is: how many times would you have to fold a piece of paper to reach the moon?

Not as many as you think!

If you were to fold a sheet of paper in half, doubling its thickness, you might have 1/16th of an inch (#1). Double it again (#2), and you'd have 1/8th of an inch (#3) – still pretty thin. But watch what happens as you keep doubling it – in just 39 doublings your humble piece of paper is thicker than the distance to the moon.

| Folds | Thickness | Units |
|-------|-----------|-------|
| 1 | 0.0625 | Inches |
| 2 | 0.125 | |
| 3 | 0.25 | |
| 4 | 0.5 | |
| 5 | 1.0 | |
| 6 | 2.0 | |
| 7 | 4.0 | |
| 8 | 8.0 | |
| 9 | 1.3 | Feet |
| 10 | 2.7 | |
| 11 | 5.3 | |
| 12 | 10.7 | |
| 13 | 21.3 | |
| 14 | 42.7 | |
| 15 | 85.3 | |
| 16 | 170.7 | |
| 17 | 341.3 | |
| 18 | 682.7 | |
| 19 | 1,365.3 | |
| 20 | 2,730.7 | |
| 21 | 1.03 | Miles |
| 22 | 2.06 | |
| 23 | 4.1 | |
| 24 | 8.3 | |
| 25 | 16.5 | |
| 26 | 33.1 | |
| 27 | 66.2 | |
| 28 | 132.4 | |
| 29 | 264.8 | |
| 30 | 529.6 | |

| Folds | Thickness | Units |
|-------|-----------|-------|
| 31 | 1,059.2 | Miles |
| 32 | 2,118.3 | Miles |
| 33 | 4,236.7 | Miles |
| 34 | 8,473.3 | Miles |
| 35 | 16,946.7 | Miles |
| 36 | 33,893.4 | Miles |
| 37 | 67,786.7 | Miles |
| 38 | 135,573.5 | Miles |
| 39 | 271,146.9 | Miles |

The distance to the moon is 238,855 miles, which means in just 39 folds we're there!

Now imagine what has happened in computers. Moore's law first "started" in 1965, which is about 27 doublings – so we'd have gone from 1/16th of an inch to over 66 miles in our paper-folding experiment. That is an unimaginable difference, and it shows in the power of computers today. In fact, modern computers have gone from a few hundred transistors to 2.9 billion in the latest Intel i7, with specialty chips reaching even greater heights. As consumers we can take these things for granted, as they show up as faster games, bigger spreadsheets, and of course phones that do almost magical things compared to what anything could manage even 10 years ago. For AI, it has meant that entirely new things are possible.

## Algorithms

The basic idea of stacking artificial neurons into a neural net is not new; in fact, Steve Jobs was trying to do this in the 1990s with his "NEXT" computers. Back then we had no idea how to deal with neural nets, and his approach was to program them versus train them.

In 2006, a researcher at Stanford University, Fei-Fei Li, created a set of data for training AI to recognize images, called ImageNet. It formed the basis of a yearly competition at Stanford, where each year different approaches were tried to do a better and better job of classifying data. And each year they got a little better.

Then Professor Geoff Hinton and his team from the University of Toronto brought a deep learning model to ImageNet, and showed a 10% improvement, blowing the doors off any previous team. As academics do, he published his approach, and the rest is history.

Almost all AI you use now has deep learning at its heart, and the two things you need to remember is that it is 95% about the data, and about the level of accuracy you require.

## Training AI

Previously, we looked at how regression is a good analogy for how we train data in supervised learning – you tag all of your data and over time the computer learns that things that look like your data are correlated with whatever the label is.

So if every time the temperature goes up by 1 degree, the humidity goes up by about 2%, you have a simple equation of 1 temperature = 2 humidity. If the temperature is 85 degrees and 50% humidity, and the temperature goes up by 5 degrees, I should have 60% humidity. Pretty simple.

Except that's not how the world really works, certainly not all the time. The problem with use of data to create these sorts of relationships is that the world is messy. So whatever relationship we can figure out will only cover some percentage of the real world, and we have ways of measuring this accuracy – in the case of regressions we call it the *R*-squared.

Possibly the most important thing you need to take from this chapter is that no training covers everything, and no computer is able to operate beyond what you've trained it to do. It is guaranteed that if you use an AI system long enough, you'll find things it cannot handle.

We train AI so it will learn to recognize and operate on real examples and real work. That means we need lots of data, and that data needs to be carefully prepared so that you're not training your AI to recognize the wrong thing.

A great example comes from a TEDx Dirigo[1] talk given by Peter Haas, a robotics researcher, in 2017. Haas tells how they'd trained a

---

[1]https://www.ted.com/talks/peter_haas_the_real_reason_to_be_afraid_of_artificial_intelligence?language=en

system to recognize animal breeds, and found that their AI misclassified a husky as a wolf. Most people would imagine this is an innocent and understandable mistake, after all they do look a lot alike.

Haas and colleagues weren't satisfied by that explanation, so they wrote some code to find out why the AI thought the husky was a wolf, and what they found was surprising. The AI wasn't looking at the eyes, or the snout, of the animal. Instead, the AI noticed that all of the photos of the wolves had snow on the ground, and the picture of the husky also had snow on the ground, so of course this animal in the snow must be a wolf.

In 2017, it is worth remembering that many thought AI was the biggest threat facing humanity, surpassing climate change, war, and disease. And Haas's message was that AI is not some super-competent technology, but it is the ultimate idiot-savant.

We need to train our data carefully, check what it comes up with, and always have a human watching because of the most important fact about AI – it is stupid.

# CHAPTER 7

# Applying Artificial Intelligence

Is artificial intelligence coming for your job? Maybe. After decades of AI progress, and about a decade of the new generation of deep learning AI applications, the promised job losses have not materialized. Like any technology, AI will change the jobs landscape, but be careful about believing the hype about how much, and how quickly AI will replace what humans do – it is much harder to apply AI technology in the field than it is to demonstrate it at a conference.

Why the concern? Because we have seen US industries like manufacturing, steel, even farming, lose huge numbers of workers in the past, and it looks like AI might be the next wave, coming for construction as well as other sectors – but is it that simple?

The US manufacturing sector has seen huge changes in the past 50 years, with steel mills, car factories, and entire sectors like textiles automated or shipped overseas. The reality is that this is a hugely complex picture, but it has produced an ingrained fear of automation in workers across industries.

From about 2015 until not too long ago, report after report predicted that huge numbers of jobs would be replaced by AI – many gleefully pointing out that white collar jobs were at risk for the first time. The poster child for this change has always been autonomous driving, which would throw millions of workers out of a job, starting with 3.5 million truck drivers.

Part of what made these predictions credible has been the success of car companies in automating certain well-defined actions that cars can take. Things like automatic parallel parking, lane changes on the highway, and automatic braking. These seemed magical, and it looked obvious that this progress would continue at about the same pace.

In this regard, software development, and especially AI, is like many projects: it is easier in the beginning, and the end – the final deployment or signoff – can be the hardest part. Why this is true of AI is important to understand, so you can ask good questions about AI applications that come your way.

Previously, we discussed how AI can *only* handle cases that are identical to the data it has already been shown. And to work, the machine learning that powers an AI application needs a great deal of these data examples. In practice, that means that you'll get a lot of easy to find data, and that will give you a working prototype that can do amazing things. But it is only a prototype that is working within the limits of that early, easier to find data set. This is why demonstrations do not indicate a working product, and why piloting is especially critical for AI-powered products.

Using the autonomous driving example, it is much easier to get many, many examples of driving down an empty road, or driving on a highway at 70 miles an hour, with most of the cars just going straight.

Because they could get tons of data on these use cases, the companies developing autonomous driving are able to show their prototypes changing lanes at 70 miles an hour on a normal highway, or parallel parking, and so on. However, these early wins did not translate into the more complex situations like getting onto the highway, merging, driving down a normal street – or anything like driving in a metro area. And that is because these situations happen in so many more ways, and there are so many more of the edge cases we introduced in Chapter 2.

There are two big points that come from this difference between early success of narrow applications, and long delays of broader applications. These are required *level of accuracy* and increasing difficulty of getting data on edge cases.

Let's go a little deeper into these issues, because they will define what you can, and cannot, achieve on the jobsite with AI.

## Level of Accuracy

Understanding how accurate you need AI to be has huge consequences for how much data you need, because the amount of data you need gets very, very high as your accuracy goes up. The Pareto principle, known as the 80/20 rule, applies here, but much worse. It's more like 99% of your data will be to get you the last 1%, making the last 1% too expensive for most applications. If it needs to be 99.9% accurate, current approaches to AI are still not feasible for most real world situations.

For example, Amazon uses AI to give you recommendations for what you might want to buy. In my experience these aren't very accurate, even though I buy a lot of books on its site. But it doesn't really matter, no one gets hurt by low accuracy of recommendations, and I still find what I'm looking for. We can say that their accuracy level doesn't need to be very high to be acceptable.

Amazon doesn't publish their accuracy for recommendations, but let's guess it's 75%. What is interesting is that Amazon has a truly amazing amount of data, and they are the world's largest provider of cloud computing, so they clearly do not lack the resources to crunch the data. The issue here is that the problem itself is too wide, and everyone's tastes are different enough that getting enough data to be more accurate than this gets harder as your desired level of accuracy rises. But perhaps Amazon doesn't try very hard to make the accuracy much higher, by doing surveys or other activities that would make the problem easier. They are at the right level of accuracy for what they want to achieve – sometimes getting a "wrong" recommendation means you try something unexpected, which for an e-commerce site is a good thing.

In contrast, Apple's FaceID is actually a security feature, so level of accuracy is a huge issue. Apple does publish this, and they estimate it is about a one in a million probability that FaceID will get it wrong, or 99.9999% accuracy. That is a high bar, and to reach it Apple uses more than just the camera, they use a laser to scan over 30,000 individual dots in your face. Apple makes the AI part of the problem as "easy" as possible by collecting so many reference points of the face, and doing it with a laser scanner instead of just an image. Those of you who have done reality capture on a jobsite will immediately grasp how much more accurate that will be.

So we have two examples, the e-commerce recommendation case where the problem is wide, but the required accuracy is low, so they don't try to make the problem easier. The contrasting example is FaceID, where the problem is purposely narrow, the required accuracy is high, and the application is built to capture a ton of information that provides high accuracy.

It is not hard to extend that thinking to a construction jobsite – how many problems can be narrow and collect a ton of data, and how many problems are always going to be broad and therefore a difficult problem?

Coming back to autonomous vehicles, once again you see that when the specific task to be completed is relatively narrow and we can capture a ton of data, the AI works well. Parking, emergency braking, and lane changing – all of these are hugely valuable, and with the first two, humans are often not very good at them. But they are a small part of overall driving.

Driving down a crowded street is not just more complex, it is loaded with potential unexpected events that might require a reaction. And that gets us to our second major point, which is the difference between improving performance and the data quantities required to achieve that performance.

## Capturing the Edges

Most things we do often require almost identical performance each time. Driving to work every day can feel like it is a song caught on auto-repeat. You pull out of the driveway, drive down the street, stop at a few stop signs, pull onto the highway, pull off the highway, a few more stop signs and you're in the parking lot.

Except once in a while, the idiot in front of you brakes for no reason. Or a soccer ball comes out of a yard to your left. Someone runs a stop sign, or the stop sign is hidden by a bush, or there is fog, or there is a car parked illegally, and so on.

The list of things that happen all the time is usually short, and because they happen all the time, it is easy to get data that represent all of these commonly recurring events. This means building an AI system that can handle all of these commonly recurring events is manageable, and was done pretty early in the autonomous vehicle development process.

For example, Waymo, the unit of Google that is developing autonomous driving and is regarded as among the most advanced in the industry, had driven about 5 million miles as of 2018. During this time, their accuracy continued to get better and better, as measured by the percentage of time the driver had to take over driving from the AI. This is known as the disengagement rate, and it continued to fall.

The problem for AI generally, and cars specifically, is that these "other" things that happen, the edge cases like random braking, stop sign running, and so forth, don't happen as often, and almost by definition there are more of them than the things that happen often. So you have a bigger list of events that happen much, much less. Think of how many times in your life you have actually seen someone run a red light. Maybe 10, 15 in your whole life. That might be 15 too many, but it is tiny compared to the many thousands of times you have stopped at a light and no one ran it.

The practical result of this is that the beginning of creating an AI project will look great – the demo will work. Crowds will cheer, journalists will write about a dramatic future of consumer benefit and worker demise. Then a few months or years go by and the progress seems to have slowed, then almost halted. At the same time, the company developing the AI will tell you they've figured out how to generate enormous amounts of data, and still the accuracy isn't getting better that fast.

This is because as you go closer and closer to 100% accuracy, the list of potential edge cases grows exponentially, but the actual number of times these edge cases will happen remains small, so you need to get out there an exponentially growing amount of time.

There are two possible responses to this situation: Get more data or make the problem smaller.

Both approaches are being tried in autonomous vehicles. In the US, Waymo continues to have thousands of vehicles on the road. With Google's AI resources they will probably crack it one day, just nowhere near as fast as hoped. Similarly, every Tesla on the road is capturing data all the time, which will amount to many millions of miles of data, so they might crack the problem sooner.

As we have seen, this approach means that autonomous vehicle technology will be a slow, feature-by-feature introduction process. More of a driver assist technology for the near future.

In contrast, bus routes, airport shuttles, and other mass transit applications are a great way to narrow the problem. You can restrict their lanes, limit them to pre-set stops, and so on. That is probably going to be how autonomous vehicles really become commonplace for the next few years.

To understand how amazing the human brain is, consider this: You drive all the time, it barely requires any of your attention. But this easy task has stumped all AI solutions that have been tried. The best funded companies in the world are just getting their demos to do a few things in the real world.

Think about that the next time you read a breathless report about future joblessness due to AI.

## Lessons About AI Accuracy

None of this means that AI isn't important, or that it won't help you in your job in the years to come. But we need to ask what we're asking AI to do, and how well defined the problem is. Let's distill this down to three questions you can ask about any given application, and whether AI will reasonably be able to help with that application:

1. How accurate does this application need the AI to be?
2. How narrowly can we define this application and have it add value?
3. How much data can we collect, and how expensive is it to gather?

### Augmenting Human Work

The question about required level of accuracy comes down to understanding whether the application directly interacts with the world, or is something that can have a "human in the loop."

Humans are involved in AI products more than is commonly reported. Many user-facing AI applications you'll see have some level of humans in the loop, whether it's one of those AI appointment setting apps, photo-recognition, or others. There is someone, often a contract worker overseas, checking the AI's work to make sure it's not producing errors.

Nothing is wrong with this, if it works and doesn't make the application too expensive to scale. Whether a human is required for an application to work is a good question to ask if a startup comes to you with some new AI tech – many early stage companies use humans to make up for the fact that they haven't yet trained their systems on enough data.

Having a human involved means that all of the image recognition, reality capture, and speech transcription applications of AI are going to have a lower level of required AI accuracy, because you'll have that extra layer of common sense to make sure errors don't creep into the process. The model doesn't have to do the really hard stuff, like handle edge cases. It just automates the common, often repeated steps.

In contrast, if the software is directly putting its output into the world, either by running a machine or sending information or analysis directly to your teams, customers, or management, there is no human to catch mistakes. So you need to ask the next question: what is the risk if the AI gets it wrong? Three examples illustrate my point:

1. *Navigating a drone vs. a backhoe.* Drones are in the air, so are in a much simpler environment than a machine on the ground. Drones are almost never going to have anything in their way they cannot see in advance, and they can move in almost any direction to avoid anything unexpected. A machine on the ground, like a backhoe, is surrounded by people, materials, and other machines that are doing unexpected things, which it can harm. The required level of accuracy for these two applications is different – a drone can have a lower acceptable level of accuracy, while a backhoe will be much higher.
2. *Data analysis vs. data visualization.* There are AI applications that can automatically run analyses for you. But without humans to gut check their conclusions, it is a bad idea for the AI system to directly publish reports, and I don't think anyone does that now. In contrast, getting an AI to visualize data, or choose color schemes doesn't create real risk. Even there you might not want to "let it loose," but the stakes are obviously lower between publishing conclusions and choosing how to visualize them.

3. *As-built models vs. reality capture.* Similar to example #2, an as-built model will get viewed by a number of key folks, so its accuracy can be very important. It will need to be checked by a human for accuracy. In contrast, the automated reality capture that is becoming common is a perfect thing for AI to handle, because it is an input into another process of design and model refinement run by people who can manage it if there are some errors.

## A Practical Guide to Narrowness

We've come across this idea of narrowness a few times in the book. Software is good at narrow, specialized tasks that humans are not, and humans are good at broad, common-sense problems that software is not. What do we mean by "narrow?"

Narrow in the case of defining an AI problem is not really about the problem itself, it is about the data you will be able to collect in the real world, and that comes down to how many ways the problem presents itself, how many ways to solve these different versions of the problem, and how many elements can be part of solving the problem. Think of it as three lists that get multiplied together to create the full list of how many types of data you'll need:

1. How many ways does the problem show up?
2. How many ways could you solve each version?
3. How many factors can be involved in real-world solutions?

For example, in our autonomous vehicle (AV) example again, we have a reasonable number of types of driving tasks, let's say 50. And we have a reasonable number of ways you could solve each problem, let's say 10 on average. Together, that's just 500 unique problem/solution combinations. So maybe 5 ways you could back out of the driveway, but 15 ways you could navigate a parking lot.

The third list is the hard part, because it is almost infinite. And because you can't ignore any possible examples, your list of combinations is impossibly large.

So we have three lists, two are reasonable and the third is not. That third list is about how much of the world's complexity you let into your application. Cars out in the world cannot control this, so the big wide world in all its splendor is a part of an autonomous car's problem, unless you limit that with the set routes, dedicated AV lanes, and other strategies we discussed earlier.

That is lesson one of any AI application: how can you limit exposure to the unlimited variation of the real world?

## Data Integrity: The Flip Side of Narrow

As important as it is to narrow your question or task to one that an AI model can handle, it is critical to design your process so that you are able to collect enough data to train the model well. This isn't only about quantity, it will also depend critically on the range of examples you include – and usually requires a little creativity on the process designers' part.

Because AI will learn from the data, and only from the data, you have to be careful about the AI learning things that aren't relevant, and it takes a little imagination to think of what these might be. For example, if you are training your system to recognize wrenches, and every 1/8 inch wrench is on a wood table, but every 1/16th inch wrench is on concrete, you can bet that the system will think that every wrench on concrete is 1/16th inch, and every wrench on wood is 1/8th inch.

Imagination will get you quite far in this case, but you'll also need to test your system and see how well it does – and here again the data you test it against must be as broad as possible.

Data integrity goes beyond just accuracy for functional reasons. This has huge ethical implications, because datasets that are not representative of the population they will be applied to will cause bias in the models they create. AI bias has already been called out as an issue for AI systems used for policing, loan applications, and especially natural language processing. This last is interesting because software developers have often used huge data sets of real language to train their models, and real language often reflects ideas and opinions that are actually not relevant to what you're trying to train, but creep in anyway.

The practical issue for construction applications of AI is that biases will mean bad outputs. It will mean that the system isn't reliable and therefore less useful. People have a tendency to believe what is written down or what pops out of an expensive software system, especially one that has come from a big name company or from the "innovation lab." As a result, it is not difficult to imagine a bias you didn't realize had been created surfacing a year later, well after the system has become "normal," and that these biased results are unquestioned because no one is checking them against reality, or even their own gut.

Keep in mind that AI is a little dumb. It needs your help.

## Data Collection

Much of the previous discussion assumes an organization is able to collect the data they need, in the quantities that a real AI development process is going to need.

How much data?

Assume that an original model will need hundreds of thousands, to millions of examples, depending on how narrow you have made the problem. For pre-trained models, you can often get away with much less, but most AI applications don't have a pre-trained model available.

It is often said that any machine learning project is 95% about the data. This is because collecting data usually takes some ingenuity, but also because existing data is almost never in the right format, has bad data, or is incomplete. Finessing data, re-collecting and getting it to work in the training system is hard work that takes dedicated time from a data scientist.

So before buying that shiny new AI product, ask the team developing it how much data they are going to need, how broad the data sets are going to need to be, and how they will test for biases both in the short term, and over time.

## Transfer Learning: Easy AI

Most companies have nowhere near enough data to train an AI system. Thankfully, huge data sets are sometimes not necessary, because models are out there that have already been trained and can do many of the things you want.

Google trained their natural language processing model, BERT, on all of the contents of Wikipedia. They have made that model available for companies to use, and with relatively low effort you are able to make BERT work for your natural language processing tasks – like perhaps recognizing specific instructions, words, and phrases regarding MEP trades. BERT makes a chatbot, or voice agent, much more capable and "natural," and works with very little training data – we got it to work with a two-page script, for example.

This is called transfer learning, and works well for language and image recognition, because there are a lot of high quality models that can be accessed. For other potential uses, first do a Google search on the problem, because anyone who has a model that can be used like this will almost certainly have published it, and might have some case studies. If that shows some results, this is the time to engage a data scientist, and if a search doesn't immediately show results, once again engage a data scientist, as they may know of examples that would work even if they're not a one-to-one match.

A word of warning, though. Even these pre-trained models need a data scientist and programmer to get working. The problem you are solving is the need for tons of data, and they do that well. We next tackle the question of how you can create, or buy, an AI application.

## Applications of AI

You can apply artificial intelligence to your workflows in three main ways:

- build it from scratch
- build an application around someone else's artificial intelligence API
- buy a product or service that has done all this for you.

Most companies will do the second and third of these options, but let's dive into all three.

### Building Your Own AI

Earlier we introduced much of what it takes to build your own model, and much of what you need to watch out for, so let's lay out in more detail what that process would look like in real life.

1. **Identify the need**: you'll only be considering AI if you've got something you want to get done, some need to be addressed. AI is appropriate if you want to automate something, whether that's recognizing common elements, guiding users through software, recognizing safety issues on the jobsite, and other functions like this.

2. **Decide on level of accuracy**: the need and real world application will drive this. Follow the steps in the previous section to engineer the problem so your required level of accuracy is as low as possible.

3. **Define the scope**: not everything will be "AI," some will just be plain old software. Try to make the AI part as narrow and focused as you can. This step should include design of the rest of the software that will deliver your AI, as well as the processes that will feed data back into the AI so it keeps learning.

4. **Define the data**: what kinds of data are needed to train the AI? How will you collect it? Who will oversee this – it should be a professional data scientist.

5. **Decide on the model**: there are often many ways that might work, and there are hundreds of models freely available. This is definitely the work of a data scientist, and it is critical that this person has solid, deep experience you have verified with past partners, as you're unlikely to be able to check their work directly.

6. **Collect data**: Sometimes you'll have lots of data already, sometimes you'll need to collect it. In many cases, creativity is needed to get the data you want. Some companies have created entire side businesses dedicated to collecting data for their primary purpose – you might decide to do something like creating an app or some other process to speed up data collection.

7. **Train the model**: This will involve lots of number crunching, tuning, and internal testing. The standard way to train a model is to use about 70% of your data to train the model, then 30% to test it. That works to make sure you've trained the model against the data you were able to collect, but doesn't protect you against incomplete or biased data.

8. **Test the model in the real world**: try the model with your own teams, then with customers. This is where edge cases and data you didn't anticipate will come to light, and can require you go through steps 5 and 6 again.

9. **Design an ongoing data and training process**: The magic of machine learning is it can continue to learn over time, and in the process improve. What is critical is that both the real world data, and the outcome, be included. So if the AI makes a bad classification, you need to include that in its ongoing training set. And this process needs to have anti-biasing safeguards that are at least as stringent as the original data collection.

10. **Finalize the product and ship**: by now you'll have some understanding of how often the model works and how often it stumbles. Using this information, you can decide whether, and how, a human will be in the loop. This human in the loop might not be permanent, as the system will improve with use.

Note that you'll need to do this same process whether you are growing a model from scratch, or if you're taking one of the "trainable" models, like BERT. It is just dramatically faster with a pre-trained model.

## Building Around an API

Google, AWS, IBM, Microsoft, and other cloud companies provide a full suite of artificial intelligence functions you can access via APIs. These APIs are usually mature, trained on enormous quantities of data, and most importantly, constantly tuned and updated for performance. Just that last part, that they are tuned for performance, can be enough reason to go with these APIs, because AI models that you build on your own can be huge, power-hungry, and tricky to get to work the way you want.

APIs from the big software companies will come with clear documentation, some level of support, and very likely a community that has worked with them, so you can "Google it" when you're stuck. To an extent, this will be true with other companies as well; for example, Houndify, the company behind the music recognition app

SoundHound, offers amazing natural language processing capabilities and a professional developer relations team.

While the specifics of what you'll build, and how you'll build it will depend on your own business and team, there are some points you need to be aware of:

1. It is usually better to use an API than build your own.
2. Try to outsource as little as possible, and limit outsourcing overseas. The team doing this should be in constant, clear communication with the business owners, and that's always better when it's your team.
3. Map out how much you will use the API. Your developer will need to be part of this so you understand how much you're doing chargeable things. Don't assume because there are really low API costs that at scale the cost will be low. It can add up, as can the other services like storage, emails, and others that usually are necessary.
4. Do not neglect UX. Include user testing explicitly in your scheduling, not just to see if it works, but to see if normal people not connected to the project know how to make it work.

You can and should go to the Google cloud platform, AWS Machine Learning/AI Services console, Microsoft Azure Cognitive Services, IBM's Watson, and others, and browse a bit. This is valuable because each of these companies has built different services that you can use, beyond just a generic image recognition or chatbot maker. For example, below is what you'll see on the Microsoft Azure Cognitive Services page, where they have a few high level options, including decisions, language, speech, vision, and web search. The "language" tab includes:

- Extract meaning from unstructured text
- Immersive Reader PREVIEW
  - Help readers of all abilities comprehend text using audio and visual cues
- Language Understanding
  - Build natural language understanding into apps, bots, and IoT devices

- QnA Maker
  - Create a conversational question and answer layer over your data
- Text Analytics
  - Detect sentiment, key phrases, and named entities
- Translator Text
  - Detect and translate more than 60 supported languages.

Just in this list they have a way to help you create Q & A, detect sentiment, and do translation. Similarly, under their "search" tab they have about 10 different search-like functions you can add, like spellcheck, video search, and more.

Under the "Speech" section, they have APIs that will transcribe audio, including when it is part of a video, translate, and detect who is speaking. So imagine if you're doing Zoom meetings and you want to create an automatic transcript – two of these APIs together mean you can get that in almost real time, with speakers identified. I have not built that application but they make it as light a lift as possible.

Google's cloud platform has often been one of the best in terms of performance, though their UI is famously difficult. Just like Microsoft, spending time on Google will show you a wide range of possibilities, from machine vision to speech to text to general classifiers. Google is generally more geared towards developers, so give yourself a little more time to poke around, because when you are ready to start building as a platform they have the most complete set of tools.

When dealing with any of these platforms make sure your software developers are really looking around at all of the cloud platforms, because they will usually try to put it all in the cloud provider they know best, which isn't necessarily the right choice. Push them to connect APIs from one provider to another if that is the right solution.

## Buying Your AI

For most companies, in most use cases, you aren't going to either build your own AI from scratch, nor build a product around an API. Instead, you will be using artificial intelligence in a pre-baked product, either from a big company like Autodesk or Procore, or a smaller startup like Smartvid.io or openspace.ai. Each of these will have very different approaches, driven as much by their background businesses as the tech itself.

Piloting is critical here, to make sure the AI really does what they say, and to understand data security in practice. If they are using humans to oversee their product's performance, they need to tell you and certify that there will be no data breaches. It is not guaranteed that your AI provider is using humans for quality control, but you should ask.

## Dealing with Startups

Anyone who has worked with a startup will agree that they are a very different animal than larger companies. They usually have not worked out their customer support model, so can either overdo it or be under-resourced, usually the latter. Startups are usually figuring out their product mix and overall market approach, so they may not have everything you want right away, but they'll often do some customization for you.

Startups are more likely to be the source of a really new idea. The founders of startups are going to be completely focused on a narrow set of features and end-user functions, so they'll go deep with you to solve your problem. As a former and likely future startup founder myself, I can say that you do get the attention of the final decision maker, and very little of the inflexibility that larger companies have trouble avoiding.

The benefits of startups, but also the extra handholding they require is why many general contractors have created innovation teams, so that the rougher edges of a software startup can be integrated into the larger operations of a contracting company.

Apart from the realities of dealing with smaller, emerging companies, it is important to understand as well that very few startups have access to large quantities of data, which means that most of them will have created their models using either open source libraries, or some of the APIs we describe earlier. This shouldn't matter, because you're really buying software that does useful things, not the "AI" itself. However, it does mean that what might look like a truly unique solution might not be, and if you like a particular company's solution, you should look around and see who else is trying the same thing. The value to you as the user and buyer is that, even if you like the existing company, you can often learn from the website, case studies, and inevitable "white papers" the other startups produce.

## AI Accuracy

Most applications of artificial intelligence report a result as if that result was chosen with complete confidence. But it is important to remember that artificial intelligence features are a result of sophisticated math, math that uses past training to estimate the likelihood that a given pattern, image, or snippet of language is one of many that it could possibly be.

AI results are a probability. Often, a high probability, but a probability nonetheless. This is not so different from the human mind, where we're very often doing a little guessing, or taking a bit of a leap to conclude things. The difference is the human mind is able to access lots of other information, like the situation, what just happened, the players, and location to inform guesses. AI usually only has one type of input, and usually only one example of that input. So it is remarkable that those guesses are as good as they are, but remember they are only guesses.

As AI advances and we begin to trust it with more and more important operations, this is key to remember – AI is only making its best guess.

## AI in Your Future

In these two chapters we have reviewed many of the limits of AI, as well as the possibilities it presents. The key takeaway is that AI is super-powered software, not emergent intelligence. As software, it is still limited in the scope of what it can handle. Within that scope, AI can do some amazing things.

Your role as construction technologist is to define that scope and the goals you're looking to achieve within that scope. This usually means what you want the AI to achieve (the goal), and what actions or software functions it will engage with to do so. The rest should be up to your partners, either internally or externally.

# CHAPTER 8

# Future Tools

The construction industry has spent the last decade gradually adopting digital technology. As that change has gone on inside the industry, external developments have dramatically increased the number of technology options that are available for future adoption. Major AEC platforms, from Trimble to Autodesk have grown their construction-related offerings, and the construction technology startup community has grown enormously, offering a much wider variety of tools than was the case not long ago. In a classic chicken and egg, more startups have led to more investment, which has led to more startups.

In the past, the most advanced technologies were exclusively for enterprise. In fact, almost all digital technologies prior to the iPhone started in enterprise. Even smartphones were more of an enterprise phenomenon, as the Blackberry was both expensive and limited.

Then in 2007, Apple introduced the iPhone. Unlike anything before it, the iPhone was visual, graphical, and applications-focused, like Apple's Macintosh computers had been for years. Apple quickly introduced an application marketplace, their App Store, as a way to get the same functionality you'd expect from applications on a desktop or laptop computer, downsized and optimized to work on smaller screens and lower power.

The importance of this shift cannot be overstated, because it created a totally new market for startups to create smaller, quickly developed applications. That, in turn, unleashed a wave of creativity

141

that we're still enjoying. The App Store created a distribution channel that meant anyone could create something and quickly get it in front of any number of users. The moral of this story is that you can prepare yourself for future construction technologies by engaging with consumer products that sometimes come first.

At the same time, technical developments like cloud computing, and an explosion of developer tools like GitHub, AWS, and the easier to create APIs, all made producing software much cheaper and easier for smaller companies. This meant that startups could try things, like construction software, without piles of money.

Some of these software startups have done very well. Procore was a startup once, for example, as was PlanGrid. The acquisition of Plan-Grid by Autodesk for $875 million in 2018 was one of those watershed moments that changes the game, by showing venture capitalists and angel investors that construction technology is a great place to invest.

Kaustubh Pandya, principal at Brick & Mortar, one of the premiere venture capitalists engaged in the built environment, described the situation in 2019 as the perfect combination of construction industry demand, startup activity, and investor interest. Rarely have we seen entrepreneurs, customers, and investors aligned at a point in time to move the industry forward. Pandya relates that many construction companies view digital transformation and technology adoption as critical to their competitiveness. Technology has gone from a nice to have to a must have.

An industry insider on the investment side, Jesse Devitte is cofounder of Borealis Ventures and Building Ventures, both of which fund early stage software companies in the built environment. As of early 2020, Jesse felt that AEC startups were still undercapitalized, at least in part because the sales process for construction and design software was especially long. Having been in the position of launching a construction software startup, I tend to agree about the need for more capital to allow more runway for sales development.

Most industries have seen consumer technologies take the lead in changing how they do business, especially phone-related technologies like apps, messaging, and photos. Construction definitely had further to go, as so much of the day-to-day work was paper-based, but the broad trends are the same across the economy.

This high level of consumer technology means that lots of what will have an impact on the jobsite of the future is in your hands right now, and relatively easy to try out. That is an important change from the past that has two big implications. The first is that everyone can get firsthand experience with many new products and technologies without spending any money; and the second implication is that it doesn't take special degrees, or much training to use most products. The general way that one cloud product interface works will teach you most of what you need to know about other cloud products. We all use the same general interface vocabulary of buttons, fields, swipes, and clicks.

However, there are some genuinely new technologies that take a little more explaining. In the sections of this chapter we'll review six technologies that will increasingly matter in the future – three that are a blend of enterprise and consumer: virtual reality, augmented reality, and high quality 3D scanning. Then we'll look at three that aren't really consumer-led, but will matter immensely: sensors, IoT, and next generation digital twins.

## Virtual Reality (VR)

Like many technologies that are coming into their own now, virtual reality is not new. In the 1960s, mirrors and big CRT televisions were used to try to create an immersive sense of "being there." Just like AI, there have been earlier attempts to create virtual reality that didn't work because the underlying technology just wasn't there.

First, let's define what we mean by virtual reality. Virtual reality uses multiple screens or projectors to create the illusion that the user is in a virtual world. VR blocks out the real world and replaces it with a virtual one. VR can be achieved with 360° video, or with computer graphics – the point is the illusion that you are somewhere you are not. It has typically been achieved with a headset, where each eye has its own screen, but there have been examples using projectors and special eyewear in a purpose-built room, called a CAVE.

Some readers will recall in the 1990s there were consumer headsets and location-based virtual reality installations at arcades and some vacation destinations. These had a number of problems, not least that the graphics were awful and they made people sick.

In the years since, the industry has realized that what made people sick had to do with how the human eye works. The center of your eye is called the "fovea," and it is where most of your highly detailed, and almost all of your color sensing happens. It's the center of your visual field, and it can detect changes that happen at something like 30 times a second. This sounds fast, and it is. But the fovea only covers the center of your attention.

It turns out that the rest of your retina works another way. Whereas the fovea has colored "cones" that see in fine detail, the peripheral vision is made up of rods that are black and white, which are there to detect motion. If you think of how human vision and attention will work in real life, that makes sense. You have motion sensors on the periphery and highly detailed color detectors in the center of your attention – so your peripheral vision picks up motion, then you turn your eyes to see if whatever caused the motion is dangerous or not.

Ever notice that you can see televisions flickering from the corner of your eye, but not when you look straight at them? The peripheral vision's enhanced motion sensing is why, and it means that VR needs to be at least 60 frames/second to avoid sickness. All modern VR is at least this level, so you're much less likely to get sick than in the past.

The other reason people got sick is we don't like to have one group of senses telling us we're moving and another group telling us we're not. This is why many people get car sick when reading, and it's why VR that has you moving through a virtual scene without actually moving makes some people feel nauseous. So we don't do that, we make sure when you need to move around you do it in jumps, or at least in a way that's under your control.

This and many other tricks have made virtual reality in 2020 a vastly better experience than anything in the past. At the same time, the price of hardware, and the quality of experiences have continued to improve dramatically. Overall, the barriers to adopting VR have fallen dramatically, which leaves the big question of what VR is actually good for, and where will that change construction in the near future?

## Why VR

Virtual reality is a unique experience, because it blocks out the world, and replaces it with something else. That means you can go anywhere, create any environment, and truly make people believe they are in this new world. That is remarkably powerful, and only just now are we starting to see VR live up to its potential. There are four things you should understand about virtual reality's effectiveness: presence, first person, flexible physics, and analytics.

## Presence Versus Immersion

We often call virtual reality immersive, because the user is immersed in the content. This is powerful, especially because VR is *completely* immersive. Designers are able to block out any evidence that the virtual world isn't real, so users are much more immersed than, say, a movie theater.

Immersion is about the technology, about the equipment. What really matters is that this complete immersion can lead to what we call "presence," where users really buy into the illusion that they're in the virtual world. This is a special phenomenon that seems counter-intuitive at first, because the user has just walked into a normal room and put on a headset. But the human mind is very good at acting as if what it sees is real, especially when there is a story to follow, or there is some hint of danger.

In New York there is a large VR/AR company called the Glimpse Group. At the time of writing, I run the unit that creates VR experiences for training – we make VR and AR for manufacturing, safety, construction, universities, and more. To quickly convince potential clients of VR's power, we often run the "Glimpse Tower" experience. In this application, the user is put into an elevator that launches you up to the 30th floor, where you are then asked to walk a short plank over that great height to replace a lightbulb. I've personally shown that experience to over 100 people, and every single one has reacted with some degree of anxiety, some even refusing to walk across the plank. People believe the illusion of being on the 30th floor, even when they're right in their own offices.

The hardware blocks the world out, and the content creates the illusion that the user is somewhere they are not, creates the "presence," which is psychological. The power of presence in virtual reality is not as tied to high-quality graphics as one might think. In fact, various studies have shown that the mind is pretty good at overcoming computer-generated graphics if there is an otherwise strong reason to believe the scene is real, or at least *could be* real. Especially when fear or anxiety are involved, but also when human connection is engaged.

As anyone who has seen a modern animated movie will understand, we can develop affection for computer-drawn characters, even when they look like a child's balloon with eyes. What these studios learned early on is that if the motion looks human, and the voice is real, we get over the rest of the character's presentation being a little less human-looking. This is even more true for models of machines in VR, where we're fine with it not being photorealistic.

Presence allows you to operate in VR as if it were real life, with the same engagement, and belief that actions and mistakes will have real consequences that performing an action or task in real life would. We can put people on top of a half-completed building and show them what working there is like. We build a mockup of the building before any work starts, and allow the owner to experience the design first hand.

## First Person

Presence is especially powerful because in virtual reality, unlike video or BIM model viewers, you *are* the action, it is happening to you. The feeling of being in the first person is often odd in video, because you cannot control where the camera goes – so when you're seeing a video through the eyes of the user, it can be difficult to process. Videos of first-person shooter games prove how odd it can be.

In virtual reality two things are different that make first person much more natural and powerful. The first is that virtual reality assumes you are the one driving the "camera" – your head movements decide what you're going to see next, just like in real life. This ability to decide what you're going to see next is a big contributor to presence, and it is part of what makes VR genuinely useful as a tool for doing work, and for training.

The second thing that is different is that VR by definition is a full 360° experience – you can turn around, look up, down, wherever. You are not outside of a scene looking in, you are *in* the scene. And that means you are the center of the action, and it is happening to you.

We know from years of experience with training and other VR experiences that we process things very differently when they can potentially impact us. Engagement does not mean involvement, as many television shows and movies engage us completely without really involving us. At no point in watching do we really think that what's happening in these engaging shows and movies will have any effect on our lives, on our own bodies. This is because we're watching the action happen through a window.

In contrast, virtual reality puts users right in the action, and though the user is never in any danger, the mind is wired to not take chances, so we respond as if the danger were to some extent real. But it's not only about fear and danger. When we're doing something in VR we learn it better than if we'd just read about it, or watched it.

This first person experience in an immersed, 3D environment has another, very important impact. We can see more of a scene, understand it much better, when we are in it. A jobsite makes much more sense when you are there, than when you just see a picture. A crane cockpit 30 feet in the air feels very different when you are in a virtual version of that cockpit than if you're watching a video. BIM models you can walk through communicate much more thoroughly than those you can just look at in 2D.

## Flexible Physics

Whereas presence and first person make virtual reality a powerful way to operate and train physical processes, the fact that virtual reality is completely computer generated means that there is nothing in there you didn't choose to include. You can make gravity only work for certain things, for example. Or you can make items stick together, float, whatever.

We have only started to understand how this can be useful, because we are a little stuck on things working like they always have out here in real life. It is a pattern that repeats for every new medium: we use the new medium just like we used the old medium, until we really understand the new thing and start to innovate.

Imagine if your pull planning meetings allowed the team to pull in full 360° videos of the jobsite to demonstrate a point, transporting the whole team to the recorded scene to see how safety is being implemented, for example.

Or if you could use voice to take notes in that meeting, and every note was automatically entered into a database that could then be sorted, edited, and collected for use in reports and other things. Those notes could also be collected into multiple walls for scheduling and percentage complete discussions, where each move of the sticky notes immediately changes the schedule, and so on.

The bending of reality to suit your process has the potential to change those processes enormously. Suddenly information is immediately available, everywhere, and in much more complete, immersive forms where that's helpful. You can have any number of voice assistants to help each participant and run the meeting, with everything recorded and edited so commitments are recorded, and translated into material orders or other process-related formats.

## Analytics

Everything in virtual reality comes through a computer, which means you can track everything. This doesn't mean it is automatically being tracked, just that you can track it if there is reason to. For example, you can track where users are looking, where they are in a scene, how long it takes to do things, how well those things get done, and so forth.

Virtual reality allows users and managers to exactly understand the user's performance, which can be much better than the looser, less exact observations that are often the best you can do outside of VR.

## Augmented Reality (AR)

Where VR puts the user into a virtual environment, AR allows the user to bring virtual objects into the real world. Where VR requires a full headset and blocks out the rest of the world, AR uses a pair of glasses, tablet, or phone to view these virtual objects, and by definition integrates with the rest of the world.

A more formal definition of AR is the use of lenses or mobile devices to overlay the real world with information and content. This content is oriented to real objects in space, either by spatial mapping, or by specific markers that the AR system recognizes. Augmented reality has also been called "mixed reality," but the difference between the two is marketing-speak, not actual technology.

Augmented reality works because a camera, or other sensor like LiDAR, finds something in the environment that allows it to "anchor" and get some level of bearings in the room.

It might seem that AR is easier than VR, so should be further along and more useful, but actually it is very hard to place virtual elements into a space that the developer didn't themselves create.

Recall in the artificial intelligence chapters we talked about edge cases. In the case of AR, some of the same complexity is at play; because AR is supposed to work anywhere, it means you're going to have a huge range of possible surfaces, lighting, and objects in the room that the computer will have to understand and integrate so it can put things in front of these objects.

Augmented reality has steadily become more sophisticated as the technology has developed across four distinct phases to date, with at least two expected in the future.

## Phase 1: Marker-based AR

Smartphone cameras scan a scene, and the phone recognizes something, like a QR code, or an image that the system has been trained to recognize. Usually what would happen then is that one image, video, or 3D image would appear in front of, or to the side of, the marker. As far back as 2013, this sort of thing was used by tech-savvy architects to show off designs in presentations.

Marker-based AR is still used because it is cheap and easy to execute. For example, Glimpse has a product called "Post-Reality," that allows users to create AR experiences quickly and easily with zero code.

## Phase 2: Computer vision/SLAM

A more sophisticated approach takes images from a tablet's or phone's camera, and applies a kind of algorithm known as "simultaneous location and mapping" (SLAM), which means it identifies flat surfaces, edges, and other features in the otherwise flat image. Through these image features the system determines where to integrate virtual objects to give an AR experience.

SLAM-based applications have been used extensively in retail, with brands like Ikea publishing smartphone apps that allow the user to pick furniture and place an image of it on the floor in their real home or office.

## Phase 3: Headsets

In 2016, Microsoft released the Hololens, an augmented reality headset that was a true marvel of technology. The Hololens uses a refinement and expansion of their Kinect technology, originally developed for Xbox. The Hololens digitally maps wherever you look in the scene, which allows you to import and place virtual objects into that scene. It saves that scan of the room, and the objects you've placed into it, so if you leave and come back, as long as you have the Hololens on you can see the virtual objects you'd placed there.

In 2018, a startup from Florida, Magic Leap, introduced their own AR headset, the Magic Leap One. It was effectively the same as the Hololens in performance and price, around $3,000. Like the Hololens, a narrow field of view significantly limits the usefulness of the Magic Leap One. The field of view is about one fifth to one quarter of a user's natural field of view, causing many first-time users to decide not to pursue the technology.

For the most part, the AR headsets that have been released as of mid-2020 were too expensive and limited for widespread adoption. A possible exception has been the Trimble Connect, which combines a hardhat with a Hololens pre-loaded with Trimble's software such as SiteVision.

The fact that the resources of Microsoft, and the very heavily funded Magic Leap, were only able to create a device with this limited field of view illustrates how hard it is to create a system that can effectively integrate virtual content with the real world. However, it is widely expected that headsets from Apple, Facebook, and others will be dramatically more user friendly within the next 2–5 years.

## Phase 4: LiDAR

The key function of all AR approaches is understanding what is in the environment, so that it can be mapped to whatever virtual elements are to be introduced.

While SLAM and other machine vision approaches can be effective, all of them are using some form for AI to interpret a flat picture. This means they need to take time to understand where the edges and surfaces are, which can negatively impact user experience. Kinect/Hololens does a little better by leveraging a number of technologies, including projecting a grid of infrared dots that it uses to map the space.

However, none of these approaches will have the accuracy and speed of LiDAR, which is an acronym for Light Distancing and Ranging, essentially using a laser to do radar. LiDAR has been a big part of how autonomous vehicle companies like Google's Waymo have been training their cars to drive, because it gives an exact picture of the world.

And of course, construction companies have been using laser scans for years, for surveying and reality capture.

The big difference now is that consumer devices have the ability to laser scan the world and know *exactly* the dimensions of that room and everything in it, and where the device is in that model of the room.

Shapr3D has already started using Apple's new LiDAR sensor in their iPad pro to scan rooms and incorporate them directly into the design process, with a precision and speed that was impossible a short time ago.

The potential this has for powerful, genuinely useful augmented reality is huge. We will see in the coming years what actual products hit the market. It is easy to imagine AR being used for automatic inventorying, for hyperaccurate viewing of a BIM model against the current state of work put in place, among many other possibilities.

These four phases roughly map out what we've seen in AR in the past few years, and each of these phases is still in market – not surprisingly the earlier phases tend to be cheaper and more accessible, while the later phases, more costly but more capable.

In the coming years, there are two very significant developments that are expected: AR glasses and the AR cloud. Which will come "first" isn't really important, as both will launch in a less capable form and improve quickly.

## Phase 5: AR Cloud

The "cloud" is really just a set of servers that allow information and software services to be accessed from anywhere. In 2017, Ori Inbar, founder of AWE and Super Ventures, put forward the idea of using the cloud to provide augmented reality services anywhere. His idea went beyond standard cloud services, though, to include a highly detailed map of the world, accessible anytime, anywhere.

The point of an AR cloud is to create a digital twin of the world, one that can be viewed through your phone, glasses, or other viewer. With an AR cloud, users can come to a place they have never been before, and by accessing the AR cloud, immediately integrate into the local economy and society – all from their phones or AR glasses.

This technology hasn't been entirely worked out yet, but everyone from Google to Magic Leap and at least one international consortium of engineers is working hard to make this happen. The potential benefits to construction are much more powerful AR apps on the jobsite, with the likelihood that creating new AR applications will become significantly easier in the future.

## Phase 6: AR Glasses

All of the AR technologies so far assume a viewing experience that requires users to stop something else and look through special head-sets, or their phones or tablets. This is powerful, and is going to get even more so in the coming years.

However, the ultimate vision for AR is that it becomes something that works all the time. The everyday AR solution that the industry has been expecting is a pair of glasses that you wear like sunglasses, and can be worn all the time. Several companies have indicated, or been rumored to be working on this, but it is a hugely difficult challenge. Three reasons why:

1. **Processing power**: Understanding a scene and then placing things in that scene takes an awful lot of number crunching. Getting all of that to sit on someone's nose is not easy – the average pair of glasses is about an ounce, while the Hololens 2 is 1.28 lbs, or over twenty times the weight of a pair of glasses.
2. **Heat**: Whatever processing power is finally used, it will generate much more heat than users are accustomed to bearing. That will mean some clever designs to keep the heat away from faces.
3. **Sunlight**: Current AR headsets don't work well in sunlight. The bright light washes out their lenses. Obviously this needs to get solved, but is harder than it sounds: too bright and it hurts users' eyes, too dim and it gets washed out. One solution is to create

"black" pixels that block light, another is to create flat colors that create images by reflecting light, but this again is very difficult. We will have to see how this gets solved as these products come to market, potentially in 2021 and beyond.

Augmented reality is going to be one of the most pervasive, important technologies on the jobsite in the coming years. For this to happen, though, it is going to need to be a widely accepted consumer product first. This will likely mean that applications of AR are going to be user friendly, designed for a short learning curve.

## 3D Scanning

Laser scanning has been part of construction for years. Firms such as Leica, Matterport, Faro, and others have amazing technology that allows for hyper accurate scans of almost any site. The software to utilize these laser-based scans has also continued to improve.

What we're seeing coming onto the market is an extension of the same computer vision technologies we covered above in the AR section – increasingly high-quality scans that are coming from consumer-grade devices. As an example, Matterport, maker of a range of scanning and photography solutions for real estate and the built environment, released a mobile app in April 2020. This app allows consumers to scan a room with very easy-to-use directions, and create a model of that room.

This is known as photogrammetry – using photos to create a 3D model of a space or object. This is the sort of thing that a laser scan usually does, except in photogrammetry you have colors and patterns, because of course it is a photo. Even the best photogrammetry will not create a model that is as accurate as a laser scan, but for many things that doesn't matter.

Companies other than Matterport have released similar scanning and photogrammetry apps, though usually not quite so well packaged. The important point to recognize is that machine vision is getting better and better at "seeing" the world, and will continue to get better.

Remember that a general rule about AI, and machine vision is a great example of practical AI, is that passing time will inevitably lead to better performance, because each of the three ingredients of AI will keep improving: we'll keep getting faster processors, more data, and better algorithms.

And as machine vision keeps getting better, it will start to reach a tipping point where it's not just better, but it's *different*. Your handheld devices, and the apps that they can support, will not just be able to make out a flat surface for AR. Applications will go beyond creating a 3D colored map of a room, and will start to identify *what* is being scanned. They'll go beyond shapes to identifying specific products, assemblies, and more. Google Lens already does this with consumer items, and is rapidly getting better.

This sort of scanning can create the inventorying capability we discussed earlier, but also can recognize danger on the jobsite and detect issues in work put in place, including variances from the approved design. In fact, this is already being done on a much less sophisticated level by companies that use machine vision to enforce social distancing on the jobsite after Covid-19.

The AI chapters outlined some limits of AI, and this sort of scanning and real-world understanding will demonstrate those limits as common things are easily identified but edge cases continue to frustrate the system. For the near future, photogrammetry and LiDAR-assisted scanning will be able to identify some things, and will get others wrong. However, the current path these technologies are on guarantees someone will create amazing applications. Perhaps even you.

## Enterprise-led Technologies

The barcode was developed in the 1950s to keep track of industrial items, finally reaching commercial success in the 1970s with the familiar supermarket barcode scanner. The need to understand where products are in warehouses and supply chains has led to all manner of technologies, from RFID to iBeacons and others. Some of these involved scanning something on the item, others worked by emitting signals from the item.

At the same time, humans have been sensing their environments for at least a century with thermostats, and using the information these sensors produce to control characteristics of those environments. We have light sensors, heat sensors, humidity sensors, motion sensors, chemical sensors, and so on.

These two different types of technology are both about understanding the environment: in the barcode and RFID case, it is about understanding the movement of things through the environment; in the sensors case, it is about understanding the state of the environment. In the past, each of these would be a silo of data, usually just one machine capturing some type of data that then went to one type of use, like counting items sold, or items in a distribution center, or keeping temperature, humidity, or light at the appropriate levels.

The networking we discussed in Chapter 3 has begun to string these sensors together, until someone made the brilliant observation that there isn't really any difference between an internet of people sending emails and other information to each other, and an internet of machines sending data to each other. So the term "Internet of Things" was coined in 1999, by an executive at Proctor & Gamble. In fact, apparently the first-known internet-connected device was a Coke machine at Carnegie-Mellon University.

The Internet of Things, or IoT, is much more than a connection of data producing sensors. The point of IoT is real-time knowledge of what is going on across the jobsite, manufacturing plant, or supply chain. That real time, detailed sensor flow produces a great deal of data, which in turn gets fed into machine-learning models. In turn, these models can predict when machines are likely to need repair, a service General Electric, Siemens, and other industrial companies have been providing aircraft engines and power plants for years.

The more accurate the data that goes into these predictive models, the more accurate and specific their predictions can be, so there has been an ongoing movement to place more and more sensors on more and more things. The benefits for supply chain and production processes have been significant. We are beginning to see some of the same approaches in construction, so it is worth a quick tour of these technologies so you understand them when they show up:

1. **RFID**. This stands for "Radio Frequency Identification" – these are often used by consumers for toll roads, for example in the US northeast as "EZ-Pass." RFID tags use a chip that absorbs radio energy from a transmitter that is relatively close by, then retransmits the unique ID of that tag. RFID is useful because

the tag itself doesn't need to store any energy, and they can be read at some distance. However, RFID tags can cost upwards of $0.10/piece, which adds significant cost over the lifetime of a process, and many RFIDs that are suitable for a construction application would be significantly more expensive.

2. **iBeacons**. Understanding where a given item is inside a space can be difficult. GPS is not very accurate, and often will not penetrate into buildings. iBeacons and Eddystone are two versions of a technology that uses Bluetooth signals to allow much more accurate location of an item within a space.

3. **NFC**. Near-Field Communication is what allows smartphones to pay at checkout without contact. NFC is often used to create the first connection between devices, which is then replaced by more powerful communications, like Bluetooth.

4. **Accelerometers**. Understanding if a damaged item was mishandled in transit can be valuable. Accelerometers are sensors that register sudden movement, like being dropped. Smartphones have accelerometers that allow the device to sense movement, but most IoT accelerometers will be tuned to look for potentially damaging events.

All of the sensors mentioned relate to moving things; there are many other sensors for rooms and fixed locations, like gas sensors, proximity sensors, motion sensors, people counters, and more.

## Knitting Together a New Picture

It is not easy to pull these sources of data together into something useful, and it's not always the case that more information is better. For example, knowing that the temperature varied by 5 degrees across a given day, from 58 to 63 degrees is a little useful. Having minute-by-minute records is usually not very useful.

What typically happens is more information leads to better questions, which in turn lead to more information gathering. But this doesn't come cheap; you need to spend the time applying intuition and judgment to these data analyses, and usually need to hire outside resources to set up and tune the system so that you're getting the information you need.

Other industries, from manufacturing to oil and mining, have found that wiring up the entire supply chain enables more accurate planning, reducing the need for excess inventories on site. Just making that work across all of the materials that go into a project will be a significant improvement.

Where the IoT truly changes the game is allowing managers at all levels to handle much greater complexity than they otherwise could, because what used to be part-guesses about how fast materials are being used, or how much is on site, or how weather has affected work, all become hard facts that can be analyzed with certainty.

The best way to keep managerial focus on the hard problems is to offload the constant checking-in on project status that can consume most of a day. This is a major reason why dashboards are so popular, and also why we see data getting graphed and visualized so often: these tools allow quick reference to the metrics that matter, and by visualizing them, managers can immediately understand changes in those measures and see where they stand against a benchmark.

## IoT and BIM

Building information models (BIM) were originally intended to represent what a building *is* and how it could perform based on its materials, machines, and design. IoT is allowing companies to expand BIM from a static picture into a dynamic understanding of what is happening in their building. With this level of connection between design, as-built, and operating data, the process of creating, and then running, a building can become much more efficient.

BIM can sometimes be referred to as "4D," or "5D." These are a play on the idea of 3D being three dimensional, with the 4th and 5th "dimensions" being cost and schedule. In recent years, this has been extended to "6D," which means using building information models to run the building over the course of its lifecycle. Since the operating phase of a building represents upwards of 90% of its lifetime cost, it makes sense to tie this operating to an accurate representation of the building and ongoing sensor-based data flows about what's happening in that building.

Construction has traditionally not really concerned itself with what happens after the building is handed off. The task of making sure that what is built satisfies the owner's operating requirements has been for the design team and the owners themselves. However, many new owners are requiring that the design, build, and operate phases be integrated more tightly, especially technology firms and large-scale real estate developers like Proctor & Gamble. This is where DFMA overlaps with technologies like IoT and BIM, as decisions made in the beginning about how to construct a building's design can have a big effect on operating costs over the decades-long building lifecycle – a somewhat obvious point that is beginning to be reflected in the information side of the process.

Other industries have found this tight relationship between what is built and the operating data it creates a huge revenue opportunity. Collecting the data, analyzing it, and providing it back to the owners of jet engines, power plants, and locomotives has become a paid service for companies like Siemens, GE, and ABB, among many others. These services provide the greatest value when they create predictions of maintenance requirements in what are called "digital twins."

## Digital Twins

In the 2000s, these big industrial firms began marketing these data and prediction services, and the term "digital twins" was coined. If you think of a BIM model that is hooked into real operating data, it will start to look a lot like the real-world machine or building it is representing. If the settings are correct, and there are sensors in the right places, the digital twin will show heat, movement, and wear and tear just like the real thing, allowing remote monitoring of the real thing.

The digital twin idea is becoming reality in more and more machines, industrial plants, and obviously data centers. It is a natural progression, from the blind building where anything you might want to know requires a middle-aged man in a faded dark blue maintenance uniform to go out and get to a real-time, flexible, and increasingly complete view of the key operating data a building creates. Sensors on pipes, in HVAC, and other building elements that might break down will provide the data that analytics platforms can use to send a crew out prior to that break down, or at least immediately after it happens.

One of the impediments to instrumenting real buildings the same way a machine can be has been the difficulty of doing so after the building is built. The other difficulty is that most sensors collect one or two data types, but a given part of a building can fail in many ways – ways that are often the fault of the occupants, not the designers. Because this is a much wider range of possible failure points, driven by the much wider range of what people do in buildings, choosing sensors that are likely to catch a problem, and thus pay for themselves, is much harder in a building than it would be in a jet engine.

This is where machine vision is going to utterly change how we instrument buildings. As an example, in 2016 we ran a technology event in the New York area – a weekend long hackathon with some of the top IoT companies in the area. This was only four years prior to this book's publication, and even so we were stumped by problems that are much easier with today's off-the-shelf machine vision services. Specifically, counting people on the sidewalk, measuring how close pedestrians got to the street, and so on, was very hard to do with proximity sensors and other motion-based sensors. In contrast, any camera is good enough to sense almost anything that is visible, because of machine vision, it is a software problem instead of a hardware problem. And software problems are always better, because we can change the software easily.

## Technology Going Forward

Companies like DPR, Mortensen, and many others have dedicated innovation teams that are constantly looking for ways to incorporate, and create, technologies and applications that make their operations more safe and efficient. In many of these cases, they do so because their customers expect innovation, and will pay for it. This is obviously not the case for many owners, so progress is uneven.

A major goal of this book, and this chapter, is to lower the perceived risk in these and other technologies by explaining what they are and what they do. Almost all technologies that are going to matter for construction have a consumer-level version that anyone can buy and start working with. While creating your own applications is likely beyond the reach, and usually beyond the interest, of most contractors, there is enough help out there in the form of unions, incubators, and analysts that without too much work, you can begin trying these and other technologies.

# CHAPTER 9

# Innovation and Technology Adoption

---

"Invention is 1% inspiration and 99% perspiration."

– Thomas Edison

---

Like many industries before it, construction is going through a massive digital transformation. This is happening because of pressures inside the construction industry, and outside of it. The internal pressures are obvious – for higher productivity, greater safety, and speed. External pressures range from new types of owners, endless innovation in technology, and the uncertain requirements that a post-Covid-19 world will place on owners, architects, and builders.

Digital transformation promises to address these pressures and in so doing advance construction as an industry. The process of transforming construction into a digitally sophisticated industry has come far enough that for most companies adopting new technologies is a competitive necessity.

The same can be said for professionals out in the field. Whether it's as a specialty trade, superintendent, or other work, digital tools are becoming part of the required toolkit. From the ground up, construction is transforming.

Transformation processes are viewed as part of innovation, and the groups tasked with managing new technology intake and implementation are usually called the "innovation" group. Innovation is a big word, so we're going to break it down a few ways and build up meaningful definitions that help you get your job done both now and in the future.

This is a chapter about three kinds of innovation that will help you take part in and drive digital transformation:

1. Product innovation
2. Internal process or usage innovation
3. Innovation to change attitudes

First, though, what do we mean by "innovation?"

---

**Innovation is creativity applied to a product or service to create greater value for the business or user.**

---

Innovation is not, itself, creativity. The two get mixed up, but there is value in keeping them separated, because they each require different processes.

Innovation is not the idea, the insight, the "solution." Innovation is the process of applying and adapting an idea for real use in the world. No idea on its own is ready to be used, there is always much more work to be done to flesh it out, and evolve it with real users in a real environment. Since so much of construction technology has to be created in an office but used in the field, the evolution part of innovation is especially important for us.

Now that we know what innovation is, let's revisit our three types of innovation that drive digital transformation: new product, new process, and corporate practice. Each of these deserves its own section, and we'll start with perhaps the easiest type of innovation to describe – creating a new product.

## Product Innovation

Product innovation can happen in a few ways: Vendors innovate new products, users create new uses for existing products, and companies internally develop products or configurations of existing products.

## Vendor Innovation

We've covered innovations and products throughout this book – it is worth reviewing how the construction professional can understand what's out there and stay up to speed without burning too much time.

Construction is a very personal industry. People like to talk to people, and this leads to there being a ton of conferences, meetups, and other places where you can learn about what's out there. Many of these will have developed a hybrid in-person/virtual model that is likely to keep adjusting in the years to come, which means you can learn a lot without traveling. Here's a good list to get you started:

### Lists and Blogs

- Blue Collar Labs – this is a listing of startups that includes a newsletter featuring up-and-coming technologies (www .bluecollarlabs.com)
- JB Knowledge and the ConTech Crew – JBKnowledge has yearly reports on technology, videos, and a weekly podcast called the ConTech Crew that is a great way to hear from companies
- Capterra is generally a good source of information for existing technologies
- Construction Dive has a constant stream of great articles about technology
- ENR is the big dog, covering the entire industry, including technology

### Union Events

Everyone from the MCAA, UA, AGC, TAUC, and others has local and national events. There are far too many to list, and many are closed to non-union members, but worth keeping an eye on regardless.

### Events and Blogs

- Autodesk University (AU) is the biggest show in AEC. A little more focused on the design side, AU has grown more field-focused in recent years.
- Procore Groundbreak is increasingly important as a place to learn about Procore, but also the many partner companies that work with Procore.

■ ENR FutureTech is an offline extension of the industry's main publication.

■ ConTech Crew, an offline extension of the popular podcast, is an event that's more like a tour, with at least major cities visited.

Events almost always charge for admission, but the blogs and videos are free. It has been my experience with all of these that they are always eager to hear from the field.

## User-led Innovation

It is very common for users to innovate by finding new ways to use old products. This is actually how modern construction technology got bootstrapped in many firms, as field personnel had become so used to digital technology in every other part of their lives, and started to apply that experience and sense of digital empowerment when faced with challenges on the jobsite.

If you've been using MS Excel at college and for your own finances – or simply got used to the simplicity of Netflix and Amazon for entertainment – the idea of running everything through paper looked crazy 10 years ago, even more so now. Earlier I referenced James Benham's comment about the consumerization of technology, and this is the process that leads to the boot-strapped innovation that's been going on in construction for at least a decade.

For example, Blake Berg, founder of DB, tells the story of when he was a superintendent at an ENR 50 construction company. He was on a crew that was doing an unusually complex refit of a data center, where three floors of servers needed to be moved into one floor, all without interrupting the data center's core operations.

This meant that there was a constant flow of design changes, especially to wiring and placement of machinery. To keep track of these changes, and to give the crews on the job immediate access to the latest, up to the minute drawings, Blake created a touch-screen, folder-centric "application" that worked on any tablet. This was not a huge technical leap, it was just an intelligent configuration and application of existing technologies to solve a problem.

Despite the technological flavor of the word "innovation," the reality is that every crew on every job is creatively solving problems on the jobsite every day. These are often smaller, one-off problems, but the fundamentals are very similar.

What makes a good trades professional is some of the same problem solving that makes a good innovator. Technology now means that the same mindset that allows a pipefitter to figure out how to solve one of the many problems that pop up on a normal job will allow him or her to innovate more broadly.

This idea of user-led innovation has a long history, and it's important for all three actors in the innovation process to understand it – users should know that they are part of a rich tradition that goes way beyond construction when they try to get products to do things beyond what their creators intended. Companies need to know how their users are actually using the product, both because they should know if they're getting value, and possibly because the company might choose to support or prohibit these new uses if they are valuable, or in rare cases dangerous. One might understand how technology can be used dangerously, apart from perhaps reckless use of machinery, but the unseen danger in everything digital is that it creates security vulnerabilities that can impact both the companies and their customers.

Finally, the software and technology vendors themselves need to know that on-the-ground users might be using their products in new ways. This form of innovation has been studied since at least the 1980s by researchers such as Eric Von Hippel, who found that many early technologies are introduced without a great understanding of how they'll actually be used. Early adopters then take the products and make changes to them that better adapt the products to real conditions.

In an industry like construction, processes in the field can be quite complex and difficult to genuinely understand from the outside. The actual work done in the field is by definition an adaptation of other people's ideas to the realities of the situation – putting work in place is, after all, figuring out how to make plans into a building. Trades and GC crews adapt things to make them work as a central required skill – so it is no surprise that technologies get adapted to make them work.

The value of this "figuring out" is very often overlooked by technology vendors focused on their own internal processes and the complexity of selling a solution to many customers who, in addition to the field adaptations that might happen, have their own office-driven changes.

One company that has made a point of listening to what's really going on in the field is DADO, a two-year old software company that found that its very first step as a company was to go out of the office and organize, any way they could, happy hours, interviews, and discussions with crews in the field. What they found stunned them, and completely changed what they thought they'd do with their product.

Jake Olsen, DADO's CEO and co-founder, tells me how they'd originally intended to create a BIM-focused solution, but upon talking to folks in the field, realized that immediate, easy, and accurate access to whatever information they needed was the big missing link in today's jobsite. What was needed by field workers was a sort of Google for the jobsite, and DADO set about building that.

But they didn't stop there – DADO constantly hosts happy hours and other small events for trades teams to keep understanding how their product gets used, and to test out new features. This discipline is rare and at the same time essential. It is not enough to just see a need in the market, it must be validated constantly. Innovation is not just about new ideas, it is primarily about the million and one decisions that are made to turn a product into a reality that works in the marketplace, and DADO's tight cooperation with real workers in the field ensures that their innovations serve their intended users much more effectively.

As Jake mentioned: "Most field workers have no idea how hungry the technologists are for their insight. We need to know what they need so we can make things that will help."

For those of you in the field, I hope to hear from you; we need your voices to help drive better innovation.

## Internal Innovation

Many companies have their own technology teams, often just a few developers who are building connections between different types of software, but sometimes more. How much more depends on the strategy of the company and how much effort they want to spend on making software and other technologies a deep part of what makes them better than competitors.

One way to tell if a company views technology as critical to their competitiveness is if they have a Chief Technology Officer (CTO). For many, the difference between a Chief Innovation Officer (CIO) and a CTO isn't clear – so let's clear it up.

A CTO should be focused on the technology the firm offers to the market, to customers outside of the company. A CIO is focused on technology that is used in the company's core operations. So the IT team sits under the CIO, whereas product teams sit under a CTO.

Construction companies are not software companies, so having a CTO is probably going to remain a rarity, but what we have seen in construction, as well as elsewhere, is that a distinct software solution that has been adapted and vetted by crews inside the company can sometimes be sold outside of the company. Personally, I think that the managerial attention that really commercializing software would require is a distraction for most companies, but it does happen. I asked Pat Sharpe, a lifelong AEC technologist, for some examples, and here's what he thought were some good ones:

- Bovis Lend Lease: Hummingbird (Image Management)
- Arup: Columbus (File Navigation)
- Leighton: Incite (Project Management) (acquired by Aconex that was then acquired by Oracle)
- T&G Constructors: RedTeam Software (Back Office)
- Katerra: Apollo (Modular Construction Software Suite)

More often, though, internal innovation is about connecting existing products so that they work the way the firm wants to work, or just connecting them at all. It has long been an issue, both inside and outside of construction software, that one software doesn't work with another, even though what they do is closely connected – for example, construction management software that doesn't connect to estimating solutions, or to the accounting system. This can also happen with file types and formats. For instance, the design team will often provide Revit-based BIM models that are not appropriate for a CNC machine, so something needs to translate them. There are commercial options, but some companies have found their own workarounds.

Internal innovation very often ties much more closely to a company-specific process and can be a huge source of competitive advantage, for at least two reasons. The first is that whatever you build will be an extension of processes you have already innovated, and that's almost impossible to copy. The second is the technology itself is unlikely to have been commercialized, and so also very hard to copy and make useful outside of the specific technologies that the firm is using.

The hard part for internal innovation is that construction companies are not software companies, so the average worker is not thinking of technology as a solution – and if they are, very often do not feel like they're in a position to do anything about it.

As we'll see in greater detail, the answer to this is to shift the culture of the company from the execution-focused, top-down approach that was the norm for many years, to one where ideas are encouraged and there is a process in place to actually make some of these ideas turn into reality.

This shift to a more bottom-up approach is done a variety of ways, from innovation contests to hackathons to design thinking sessions. All of them help, not because any individual activity like a contest or hackathon is going to produce a winning idea, but because over time they shift the culture towards one that embraces technology as tools that they can create, that they can own, instead of tone-deaf software that is a pain to use and doesn't really do what it needs to.

## Innovation as Culture Change

There is often a sense in construction that "we've always done it this way," so new processes have at least as much, but often more, resistance to innovation than products do. And as you might expect, many technology vendors express frustration at this perceived lack of willingness to try new things.

First, "always" usually means about 25–30 years, or the length of time the average construction professional has been in the market. New generations of workers always cause change in their workplace, and that is definitely happening in construction.

In fact, before we dive into process-based innovation, let's spend a minute thinking about how open to, or actually eager for these changes the workforce of the near future is likely to be, for two related reasons: demographic changes and crisis-driven changes.

## Demographic Changes

It has been said that "demography is destiny." For all we talk about technology, it is people that make an industry, and people define the character of that industry. As new generations come and go, they bring with them changes. Let's look at what that means.

Marking out distinct generations is not science, because we're having new crops of kids every year. However, the value of a bracketed generation, like the Baby Boomers, is that they all will have experienced many of the same things and thus have a similar outlook on life. Because a generation implies one cycle of birth, maturity, and birth, generations are usually 20 years, which makes them a little less useful as a way to describe a group of people, because of course some of what were formative experiences for the older, earlier born members of a generation will have happened before the younger, later members were even born. For example, the Baby Boomer generation is anyone born from 1946 through 1964. This is because in the US and elsewhere, soldiers came home from World War II and had lots of babies. The generations as generally understood are:

1946–1964: Baby Boomers
1965–1980: Generation X
1981–2000: Millennials
2001–2020: Gen Z

You'll note that Gen X is a little shorter than the others – that is in part because Gen X was characterized by low birth rates, which started to change around 1980. As a result, Gen X is the smallest generation both because of lower births per year, and fewer years. As I said, it's not really a science.

The Boomers define the current state of the construction industry. They are responsible for taking the industry they joined in the 1970s and 80s and evolving it with the technologies and legal environment of the day into what we see today.

Gen X, already a small generation, were disincentivized to stay in construction during the 2008 financial crisis, leading to a serious talent shortage. Given that the youngest of Gen X is 40 at the time of writing, it is unlikely that they'll be part of solving the talent shortage.

Millennials saw the 2000 crash, 9/11, student debt levels that dwarf their parents', the 2008 crisis, and now all of the 2020 crises. It is the most diverse generation of any large country in history. Much like the Boomers, who demanded change in the 1960s, Millennials are demanding change in their work lives, and this is showing up in their limited participation in construction.

As many of you know, the average age of a construction employee is 47, meaning that the millennials have not taken their place in construction, as of 2020. Prior to the pandemic, this was leading to many construction companies to question how they do things, to look at the culture and processes that they'd come to rely on.

Most of the focus is rightly on collaboration and a culture that is famously gruff – something Millennials are generally less willing to accept. But equally important, Millennials don't just expect, they *require* a digital workplace.

The demographic story, then, is that Millennials will form the bulk of the construction industry going forward, and they are already demanding a more inclusive, less abrasive, and more technologically integrated jobsite. It is highly likely that companies who incorporate offsite fabrication and industrialized construction more generally are going to find they are able to attract more workers at lower costs than those that do not.

Millennials will drive just as much change as their Boomer parents, and that will start with a need to change processes in construction.

## Crisis-driven Change

Prior to the 2020 pandemic, there was a general sense that, to be effective, teams needed to be working together. And 20 years ago, that would have been true, as the telephone is not a great way to communicate complex ideas and a terrible way to collaborate.

In the first half of 2020, the world economy locked down and within days about two-thirds of the US economy switched from in-person to on-Zoom. This same thing happened around the world, with different countries following a similar trajectory at different times and in their own ways.

Suddenly technologies that were useful but not central became central to how we do things – some of us practically lived on Zoom, Google Hangouts, or Microsoft Teams. Meetings that formerly had to be in person were adjusted to being virtual. And technologies like

Holobuilder, Openspace.ai, and others that allow remote surveying of the jobsite suddenly became much more interesting, and of course anyone with any technology that can help with remote communication added features in record time.

At the time of publication, the Covid-19 crisis had not run its course, as neither a vaccine nor effective treatment had yet been made available, but however the crisis ends, it lasted long enough to have shocked the system enough that processes generally will be much more open to change and evolution than would have been true in the past.

As the decade of 2020–2030 unfolds, we cannot know quite what will happen or what processes will evolve. What we can confidently predict is that the construction industry will be more open to radical rethinking of how things get done than it has in many years, if not ever. There is a perfect storm of ever-increasing building requirements, a shrinking and demanding workforce, and radically challenged assumptions about how to get work done that combine to create potential for real change.

## Process Innovation

Many sectors of the economy have had their core processes redesigned in recent years. Manufacturing, hospital operations, finance and banking, advertising – it is a big list. All have changed because of the presence of technology and underlying shifts in their marketplaces, including increasing sophistication on the part of managers.

In fact, from the 1980s through the end of the last century, a steady flow of ideas washed over the business landscape, leaving radically different offices, cultures, and processes in their wake. While any change like that is complicated, one of the biggest drivers was the flexibility that digitization brings. In the past, most information was locked in paper, or wasn't really collected at all. Then a new kind of software came along, and changed supply chains, manufacturing, finance, and HR – it was called "Enterprise Resource Planning," and its effect was to make information free of any given media. It could be tabulated, analyzed, and sent anywhere to be used for decision making about how things got done.

At the same time, the business world was recovering from the shock that the Japanese economy had given the western world – processes not only mattered, but could be utterly different from what had been done before, with almost miraculous results. The Toyota Production System (TPS), which later evolved into Lean manufacturing, Six Sigma, and other process re-engineering models, all were able to re-focus companies in ways that drove up productivity, in some cases radically.

In other words, technology had come along to liberate data and information from the paper and interpersonal interactions that had locked it in place, just as an external shock had created the realization that working another way was possible, and often preferable.

We cannot know what will happen next, as the construction industry is so diffuse that change is hard, but the stage is certainly set for process redesign, especially at the ENR 400.

How do we redesign processes? How do we innovate new ways of organizing people, organizing their efforts, and coordinating their activities across a supply chain?

Let's take a detour and look at the foundations of innovation – the thinking processes that can generate ideas that, in turn, give us innovation.

## Innovation Versus Creativity

Design companies and advertising agencies have been selling their abilities to come up with ideas for a hundred years. These ideas needed to solve problems, often badly defined problems. And those companies needed to be able to come up with a series of good ideas, because clients have a bad habit of not liking the first thing you show them. So, these creative companies evolved tried-and-tested methods for coming up with ideas, for thinking creatively and producing the ideas that can change the world.

Good ideas are hard. "Writers' block" is not acceptable when a deadline is coming – so in the early part of the twentieth century creativity was studied, to the point where we began to understand how it works. This was necessary because being able to pin down how something works means you can design a process for doing it when needed.

We've got a definition of creativity that is clearly related to innovation, but is about ideas, not application:

---

**Creativity is the combination of two or more ideas in a way that is both new and useful.**

---

Just like innovation, ideas need to add value to be considered creative, but the key idea here is this combining of two other, smaller ideas. That means you need to have a bunch of smaller, older ideas in your head already in order to combine them usefully. The first part of creativity is hard work, studying and thinking and arguing until you have filled your head with facts, parts of ideas, and some sense of how things work in the world of your problem.

The thing is, our brains do not like jumbled up facts that don't fit together, so our subconscious works to try to fit them together, to find connections that make sense of it all. We can think of ideas like nodes in a network, connected to other ideas, and the more time you spend on all these ideas, thinking about them, and trying to connect them, the more you try to make them fit together so that a new solution comes out, and the better you are preparing your brain to either make that connection right there, when you're thinking, or later when you don't expect it.

And for many, their best ideas come when they're not expecting them. This is not just for advertising or design creatives. Researchers have looked at how physicists, writers, business strategists, and engineers come up with their best ideas, and the pattern is the same. There is a generally accepted five-step process for coming up with new ideas:

1. Define the problem
2. Learn all you can about the subjects the problem involves, and try hard to solve the problem
3. Go away from the problem – sleep on it, go for a walk, whatever
4. Come back, and try to solve the problem again
5. Refine the idea

This sounds like a flaky way to go about it, especially step #3. But that is actually where the magic happens. Our brains tend to keep working the problem even when we're not actively trying –

in fact, it is exactly the relaxation of your mind that allows new and unexpected connections to be created and emerge into view.

If you've ever had an idea in the shower, or on a dinner date, or noticed that new insights often come when you're on vacation, that's all the same process. Sometimes it's better to take a break, have a beer, and talk about anything else so your brain can do the work.

Creativity is an important part of innovation, and as experts in some part of the construction process, every person on the job has the capacity to innovate some new solution, some new approach to the problem.

## Lateral Thinking

A related way of thinking about creativity is what Edward de Bono, a creativity researcher, called "lateral thinking." When you're being told to "think outside the box," this is what they mean.

Lateral thinking means looking at the problem in a larger context, to find solutions that aren't obvious if you stick to the solutions and approaches you are used to. In fact, let's use the "out of the box," example to illustrate what lateral thinking means.

The phrase "thinking out of the box" doesn't originally have anything to do with an actual box that you as the thinker are stuck in; it refers to a puzzle. This puzzle has been around for a while, and I'd like you to try the puzzle before turning the page to get the answer.

The puzzle is known as the nine-dot-problem, and it is simple: How would you connect all nine dots in this matrix with only four strokes, and never lift your pen?

●        ●        ●


●        ●        ●


●        ●        ●

Most people try drawing lines from one corner to the other, or along the sides, and find they cannot seem to make it work. They are stuck with the idea of square, or "box." To solve this, it is important to ignore the obvious pattern of the dots as a "box" and just look at them as dots to be connected.

What if the lines that connect these dots were to extend outside of the "box?" Then it would look like this:

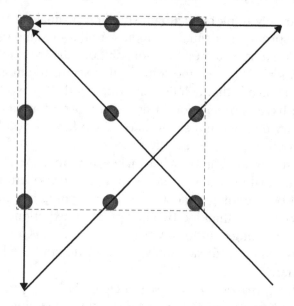

Lateral thinking, then, is expanding the problem so you can find other resources that can help you solve it. This example shows us that what keeps us from lateral thinking, from thinking outside of the box, is very often assumptions we make about how things are supposed to work. We make assumptions about what is possible, permissible, or most effective, and very often those assumptions turn out to be wrong. Especially in an industry where technology is constantly changing, assumptions about what can be done get outdated quickly.

Lateral thinking as an idea can help us realize we're probably thinking too narrowly and need to open up what we're considering, by breaking through our assumptions.

So how do we break these assumptions? Perhaps the best in the world at questioning assumptions and thinking outside of the box is

Elon Musk. He is famous for redefining industries, from payments to electric cars to spacecraft and more.

Musk does this through a radical commitment to first principles thinking, which is how you both get past assumptions and then build up new ways of thinking once you've done so.

## First Principles Thinking

Most people only think by analogy. Thinking by analogy means you look at your problem or situation, then search for examples in memory that look like it. You then look for solutions that worked for these similar situations. You pick the solution that seems the most likely to work, then simulate that solution in your mind. If it works, you're done – you have your solution. If not, you move on to another possibility and try that out. If it seems like it'll work, now you have your solution, and if not, just repeat the process.

Reasoning by analogy is what makes you an expert, it is what allows you to do things quickly, do them well, and spot errors quickly. You should be spending 99% of your life reasoning by analogy; the people who are not thinking by analogy are inexperienced and are trying to figure things out from what they see. This is a terrible way to do things someone already knows how to do, and is why training is so important.

In contrast, reasoning by first principles is hard, probably best done only when you're trying to figure out something important, and best done in a group. The process can be thought of as fighting your way down a ladder of assumptions, then building up your plan or process from the ground.

There are four stages of first principles thinking: define your goal; eliminate assumptions; define your first principles; then use "if, therefore" logic to build up a solution. This process looks like this:

1. **Define your goal.** This sounds easy, but most of the time, the first statement of your goal includes some implied ways to solve the problem. In fact, most problems are defined too close to the solution, when what you really solve is actually a higher-level goal.

   For example, "we need to get over 90% of our daily reports in." Daily reports are unlikely to be a good goal, because no one actually needs daily reports. You need the data daily reports collect, and you need to be able to prove that data is valid, that it reflects the reality it is meant to represent. Let's restate the goal as "we need to get data from 90% or more of our onsite workers about when they were there, what they did, and any issues that came up."

2. **Eliminate assumptions**. This is not easy, and again should probably be done as a team, because people are universally better at finding fault in the thinking of others than they are at realizing their own shortcuts and baked-in assumptions.

   Extending the daily reports goal, let's look at what might be assumed in there:
   - You need to collect this *from* them
   - You need to present this for some future lawsuit
   - You only need this information
   - You will only use this information for future disputes
   - This information needs to be collected as a separate exercise
   - This information needs to be collected actively by the foreman and superintendent
   - The information will be 100% accurate, and isn't useful if it's not 100% accurate.

3. **Define first principles**. Once you've got as many assumptions removed as possible, what are you left with? Is that enough to solve your problem? Very often, the stripped-down principles you are left with open up new possibilities, new arenas you might include in your solution. For example, what if your problem now looked like this:

---

"We need to create an ongoing, real time listing of who is on site, where they are, and what they are doing that is at least $x$% complete and $x$% accurate."

---

How did I get to that? Think of what was assumed by end of day listing of who's there – it assumes you cannot understand minute by minute. It assumes you wouldn't have a use for better data. And it assumes you couldn't do better. It also assumes that you cannot work with data that's a probability. These don't need to be true.

4. **Build up a solution**. By opening up our problem down to its bare bones, we can start asking about ways to know what's going on at different intervals, or constantly. Could we install a tracker on hardhats? That's been done. Could we install cameras that employ facial recognition? That's been done. If we used facial recognition, but it sometimes was dark, or you couldn't quite confirm who was somewhere, would that still be ok? What about privacy? What about camera coverage?

A solution to this might be to install simple 360° cameras at key worksites, and a standard video camera at entrances, leverage AI and build a complete understanding of what teams are where, and correlate that to percent complete. Or it might be something else; the point is that by removing assumptions, we're able to think laterally and out of the box, from first principles.

First principles thinking is a powerful way to look at processes in a fresh way, identify limiting assumptions, and create new solutions. If there is a weakness, first principles thinking doesn't really provide us with a way to actually solve problems. The focus is on stripping away assumptions, but we need some methodology for building up whatever solution we need.

In the case of process innovation, we are going to be working with how people deal with each other, how they use tools together, and what people think about existing ways of doing things. The best tool for this sort of problem is design thinking, which we explore next.

## Design Thinking

Over time, designers and innovators have expanded on the creativity process, and developed what we know as "design thinking." Design thinking follows some of the same basic ideas as first principles thinking, because it is focused first and foremost on avoiding assumptions as you create your solution – in this case, those assumptions are what happens when ideas are just talk and drawings. The assumptions are that the people working on the problem all understand what the talk and drawings mean, when we know they usually don't. In design thinking, the team creates concrete prototypes and iterates with real user feedback early and often. In every case, we are getting clear, real information instead of assumptions and generalizations, and that leads to better solutions.

The steps of design thinking are: empathize, define, ideate, prototype, and test. These specific steps come from many years of companies creating solutions that don't actually work for people. Those failures led to important lessons about how to start the right way, keep development on track, and make the most of how humans *really* think out of the box.

1. **Empathize**: It sounds obvious but to get something useful built you need to understand the problem and how your solution will actually get used. This means more than just understanding what function needs to be performed, it means understanding who will perform it, and how they'll perform it.

   This step overlaps with First Principles Thinking. The process is often facilitated by an expert who will ask hard questions about assumptions being made as data is collected about the real customer need. At least as importantly, the moderator and team overall need to insist on real user input, not just brainstorms of what they *might* need.

   We call this step "empathize" because it is about understanding the user. There aren't many industries where the gap between office knowledge and field understanding can be so wide as construction, so every successful innovation team I've met makes this part, the understanding of the field, their primary focus.

2. **Define**: After the critical first step of listening and understanding, the problem to be solved needs to be refined and turned into something people can actually work on. Take the insights from your understanding and create concrete requirements. The trick here is to define the problem tightly but avoid starting to solve the problem by suggesting approaches.

   Think of at least these five categories of requirements:

   a. What is the current problem – what needs to change

   b. What does the solution need to produce – what is the outcome

   c. What does the solution need to consume – what are any inputs

   **d.** Where does the solution need to operate – where does it get used

   **e.** Who does the solution need to work for – who will use it

     Extending our example of daily reports, we can define the problem as:

**a.** Change: interruption and lack of understanding

**b.** Outcome: daily reports that are consistent and complete

**c.** Inputs: information from superintendent and foreman

**d.** Usage context: CM software API

**e.** User: project manager, construction manager, finance

    So, the definition of the problem is that we need to find a means of getting information from superintendents and foremen to their project managers, via their construction management software's API, in a way that minimizes their interruption, and makes clear the importance of the information.

    This process must also generate criteria. The steps mentioned will each help to do this, and you should agree as a group that your criteria are sufficient, that you have covered all necessary requirements but no more. That last part is important – really argue over whether any of your requirements are unnecessary, because you'll need them later. Extra requirements could disqualify otherwise good ideas, and cause you to need to revise your requirements later in the process.

3. **Ideate**: Ideation is the generation of ideas to solve the problem. This is often best done in "rounds" of thinking, where teams share ideas, discuss strengths and weaknesses, and try to formulate a solution.

    Brainstorms have been the methodology of choice for this kind of ideation for decades, because they are easy and respect one key idea – generating ideas isn't the same process as judging ideas, and the two processes should be separated.

    Whether you call it a brainstorm or just a meeting, the ideation step needs to be structured so that there is time to just generate a ton of ideas, then a separate step to cull those ideas and review the remaining ones, and repeat.

Having run many of these, a few pointers that might help with productivity:

– People often have a very productive first 10 minutes, as they just unload ideas they've been kicking around for a while. These are very often good ideas.
– People are often lazy, and will stop after a few ideas. Giving them a high number of required ideas forces them to really dig in and think – 20 is a number I've used in the past, because no one walks around with 20 ideas about anything, so you're sure to have participants pushing into new areas of thinking.
– Remember that creativity is about connecting ideas in new ways. So, provide opportunities for participants to share ideas. This is where sticky notes are often used. One creativity consultant I know, Bryan Mattimore, likes to have participants walk from one area full of sticky note ideas to another, so they can bring ideas together.
– End the process with some form of voting for which idea should win. Make sure that you include a mechanism for including the criteria in the voting process. People tend to forget some of the criteria, and you'll wind up having to go back and refilter if you're not keeping the criteria front and center during the culling process.

The goal of the ideation process is to have as many finalists as you are able to prototype – you want as many real options in your hands as possible, depending on what type of solution you're looking to develop.

4. **Prototype:** Design thinking's signature contribution to creative processes and innovation more generally has to be the prototyping phase. The ideas that have been selected have undergone all the "traditional" creative steps, but now it's time to break through the layers of guessing what the idea would look like and get physical.

Obviously, it is rarely possible to really build something, whether it's a new process or software, or whatever. There are many, many prototyping options that can narrow the gap between reality and your idea statement from step #3. This could be drawing up an org chart, doing role play, or writing a script of how things might work.

The reason prototyping is so powerful is because ideas are just abstractions, they are stick figures that have had all the flesh and sharp angles removed or ignored. Real life is all sharp angles and unexpected consequences. When you prototype, you have to make decisions about translating the abstract into the concrete, and that teaches you things about your idea. It is very often the case that the act of prototyping changes the idea, makes it better, makes it more likely to work.

Sometimes just prototyping an idea will kill it. It's not hard to imagine re-thinking a process, then acting it out and realizing that no one would really do the process, or that a given step is too difficult.

5. **Test**: Prototyping is an internal step. Once you've prototyped and played around enough to have the idea represented well, get someone else to try it. You do this because they will find things you didn't think of, both good and bad.

Sometimes you'll end here, but most often you're going to have learned new things that will make you change your problem statement, or new criteria, and add new ideas. If so, time to do it all again.

Process improvement is going to be the highest value-added activity you do in the next 10 years, so understanding how to come up with new processes is critical. Every technology you adopt will require a new process to support it, and a change in existing processes to make the most of it. Hopefully, you will also have a process for capturing unexpected ways of using that technology.

Learning the value of thinking beyond boundaries, laterally, of breaking assumptions and thinking from first principles is, on its own, a competitive advantage. Training your teams, at every level, in design thinking is another competitive advantage, because it instills a skepticism of assumptions and a real process for doing something about it.

## Innovation as Corporate Practice

What makes a company innovative? Let's get back to our definition of innovation again: "Innovation is creativity applied to a product or service to create greater value."

In the context of a company, that means creatively thinking about how the company does what it does, continually creating greater value. Notice that the definition does not say "greater value for the customer." This is because many corporate innovations reduce costs, improve safety, and provide other benefits that the customer might not see.

Thus far, we've reviewed creativity and processes for creating new ideas and testing them out. The key missing pieces are integrating new ideas into the fabric of the company, and making them add value at scale.

How do we do that? How do we get dozens or hundreds of superintendents to use new ideas, new processes, and new products?

First, let's look at what we know about product adoption, because even though field personnel work for the company, everyone knows that they have a lot of discretion about what tools they use, and especially how well they use them.

We have been studying how new products get adopted since at least the 1950s, when researchers in Iowa looked at the process through which farmers started using a hybrid seed corn that was superior to the normal seed.

There is a general sense that adoption of anything happens in stages, in a process, which is obviously true unless the company mandates it, and even then there are stragglers.

But what really matters is that there are going to be distinct groups – there was a book published a few decades ago that proposed five groups, and even suggested what percentage of a given population each group represents. There is exactly zero evidence that these percentages are real, so don't worry about them – and there is similarly zero evidence that there are really five groups of adopters. The groups were developed from a statistical convenience that is not important. The commonly understood groups are "innovators," "early adopters," "early majority," "late majority," and "laggards." I mention it here because you might have heard of some of this.

In my experience, what is definitely useful is thinking of three groups, because their adoption process will be different in an actionable way. There are early adopters, majority, and laggards.

The early adopters are open to new ideas, like learning about new things and will be tolerant of things not working perfectly. They will often be somewhat more technically adept than their peers, and are usually the source of the user-led innovation I mentioned earlier. Early adopters do not take very much convincing, and they don't need to see anyone else using the product. The best way to incentivize early adopters is to respect their interest and energy and ask their help in making whatever adjustments or tweaks are needed to adapt the product to your company.

The majority are not interested in being innovative, they just want better results. There is a risk/reward calculation here that is what defines them as the majority – people who don't care about the technology but do care about results.

We have found a few tips that help with convincing the majority to adopt:

1. People tend to adopt things that they see others adopting, because it lowers the perceived risk. What's interesting is that research has shown that it's not the number of other people we've seen adopting something that matters, it is the proportion. If I'm in a room with four other people and three are using a new software product, it will have a bigger impact than seeing 10 people in a room of 30.
   - The practical implication is to try to introduce new ideas in geographic or functional concentrations versus trying to introduce it in a thin layer across the company. When you've got all of one department using the new product or process, you will have the testimonials and "social proof" to introduce it to an adjacent group.
2. The product needs to feel more solid when it goes to the majority. There is a phenomenon known as the "chasm" between the early adopters and the majority, and that's mostly about a specific kind of risk. The early adopters are ok spending time figuring out how to use the new product, but the majority is not. It is actually quite rare that people genuinely think a new

product won't work but they very often think it won't work for them, or will be hard to figure out, wasting their time and making them look stupid. So, when going from early adopter to majority, be sure you've got training ready, and any bugs worked out.

The laggards are not interested in change, regardless of the risk/reward. They are at work to work, and like the certainty of doing things the same way over and over. In many cases, work is just a job and they're not interested in growing or learning. Laggards will adopt for three reasons: because not doing so is making them feel silly, they need to adopt the new tech to work with other people, or management has forced them.

Each of these three groups can be different sizes, and activities like the hackathons and contests discussed earlier will help to grow the early adopters and shrink the laggards somewhat.

Most important is that people who are not early adopters see these three things:

1. Lots of other people are using the new product.
2. People who are like them are using it. This is especially true about age, as Millennials will generally have a comfort level with some technologies that Boomers do not. Boomers need to see Boomers using something new.
3. The process of adopting was easy and well supported with training.

Consumer markets have tried the idea of an "influencer" as a way to get people to buy or adopt a new product, and that has never been proven to actually work, though it can feel easier to just get one person to be your representative. The problem with influencers or champions is that they only know so many people, so it is fine that they help, and obviously early adopters should be cultivated and leveraged for a bigger roll out, but that is no substitute for a thoughtful, concerted campaign that includes the three elements discussed previously (see lots of others, like them, well supported by training).

Adoption of a new product or process is change management, and requires an internal campaign, just like other big corporate strategic change.

## Innovation Teams

Larger general contractors and specialty contractors have created innovation teams in the past decade, dedicated to digital transformation, though that's not always what they call it. These teams typically go through three stages:

- Stage 1: The CEO creates the team and gives them some budget. They hire for the team, and spend about a year designing their approach. This stage usually involves drones and some other cutting edge tech that doesn't wind up getting used.
- Stage 2: The team refocuses on pulling in solutions they think the company needs, often going out to find champions in the field. Sometimes this works, often it does not.
- Stage 3: The team by now has evolved outreach activities, like the contests and hackathons. They have found early adopters, superintendents, or foremen who are constantly looking for an edge, for a way to improve. They create feedback and suggestion mechanisms to get ideas from the field, and a formalized way of learning what software and other technologies are out there. The key to success in this step is that the innovation team is responding to needs in the field, rather than pushing out ideas that haven't been asked for.

Innovation is key to the health and growth of construction. Companies that embrace innovation as a core competence and set of processes, instead of an outcome to be hoped for, will be the winners in tomorrow's construction industry.

# CHAPTER 10

# The Digital Construction Mindset

---

"The future is already here, it's just unevenly distributed."

– William Gibson

---

P redictions about the future of technology have a bad tendency to be all or nothing. People like to say something will "disrupt," or that it is "dead." Nothing really works like that. The future of construction technology is going to be both more and less amazing than anyone can predict, because no one can predict what will be invented, what surprising effect in the world will lead to new possibilities. We also cannot predict what brilliant technologies will fail because they are launched badly or ill-funded.

The construction industry today is enormously complex, and comprised of many segments. This will not change in the future, though I do think there will be some level of consolidation. In this final chapter, we have the opportunity to revisit the mindset discussed in Chapter 1, and use this as a lens through which to view changes that are likely in the industry, and how that will affect the large, sophisticated firms, as well as smaller, more focused companies.

## Adopting the Digital Construction Mindset

Technology augments human capability, because we separate intuitive decisions from digital decisions, with each supporting each other depending upon context. Let's look at how each will evolve over time.

### Intuitive, Human Skills Arena

In 2035, there will still be an acute skills shortage. This will be driven by demand, demographics, and technological change:

- **Demand:** The world is going to continue to need enormous amounts of construction, both to house a population that will still be growing, especially in Africa, but also to refurbish and deconstruct buildings as we radically rethink how they interact with the environment across their lifecycles. Specifically, it is likely that energy consumption in the built environment will be a focus for regulation and other incentives, leading many small and mid-sized construction companies to sell lucrative services retrofitting old buildings. Retrofitting and demolition are both highly irregular project types, and will therefore require more human oversight than other projects of similar size and complexity.

  Climate change will imperil hundreds of coastal cities and towns, leading to construction projects that range from building of walls and levees to further retrofitting of buildings to withstand flooding, to demolition and rebuilding – demolition that itself will be subject to new regulations.

  In the US and many other countries, there is also a fundamental need for infrastructure replacement and upgrading that will not be completed soon; in fact, it is likely to become either a constantly higher level of activity, or a periodic investment.

- **Demographics:** Country population profiles are driven by births, measured as live births per woman. The replacement rate for this measure, where a country's population remains stable, is 2.1 live births per woman. In almost no advanced economy is this measure above 2.0 right now, and it is forecast to continue to either decline or stay at current rates indefinitely. There simply will not be enough people to sustain current levels of economic activity in 2035, at least not as we produce those levels currently.

■ **Technological change:** In the US economy overall, skills become obsolete in about 5–10 years. This is because technological change across industries continuously leads to improvements in productivity, safety, and quality; change never stops. There is always a new tool or skill that, once introduced, soon becomes expected of everyone.

Construction has largely been immune to this, although in 2020 that is already starting to change, as more digital tools are added to the toolkit. The point here is that, as required skills change, there will always be a lag between how many skilled workers are needed and how many workers have that skill.

## Learning Technology

Because there will continue to be an overall skills shortage, we can expect that learning will become a core competence of construction companies, whether they be general contractors, or specialty trades, or other companies. This change has already taken root in the consumer goods, software, and other industries, for the same reason – they don't have enough highly skilled workers, and college is a terrible producer of many skills – so these companies have become quite sophisticated at training large numbers of staff. And doing that training on an ongoing basis as skills continually evolve.

Learning technologies are often grouped into two categories – training and performance support. Training can be anything from a simulation to a classroom to online videos. My guess is most training in the coming years will be in virtual reality, or whatever replaces it – the technology is already very good, and is going to continue to improve as artificial intelligence and other supporting technologies progress.

Simulations specifically will be fine-tuned to teach the human operator the kinds of context-heavy skills that only humans are good at, by putting them through many scenarios, quickly. This sort of training is being used now for retail management and other jobs where the ability to recognize a pattern and quickly respond is highly valued. We can expect that this sort of simulation will get developed in the construction trades and GCs, so that new team members can get trained up more quickly than traditional methods alone. There will still be a need for the apprentice to journeyman process, but it can be augmented significantly by simulation training.

Performance support involves providing information right where and when it is needed. Right now this is often difficult, as it is hard to know what is necessary, when. But as augmented reality and other technologies mature, we can expect that an AI agent or something similar will be able to provide detailed information right at the point of need – indeed it will likely be pretty sophisticated information.

Both of these modes of learning will develop the human skills needed in the future, and the reader will be able to imagine most of the content that will be learned. Outside of the standard topics and skills for a given trade, however, there will be a whole range of new skills that we have not yet developed – skills that grow out of data analytics. As software and digital construction technology continue to become more sophisticated, and as we gather more and better information from the jobsite, supply chain, and broader world, the ability to manage that data, organize useful analyses, and act on those analyses will be critical.

In short, the future of construction will be one of empowerment and organized learning, where the field worker is given the training and onsite performance to do their job safely and at productivity levels that would seem like magic today – we know this because the same trajectory has happened in other industries.

## Digital Skills

The second half of our new mindset relates to thinking of construction in digital terms. Construction has always involved a constant flow of information, and that flow has now become digitized. This book, courses like my Procore.org series on "data in construction," and others all offer a relatively painless introduction to the digital skills that will be necessary.

What we mean when we say digital construction mindset, and the final point I would like to make in this book, is that technology is just a tool. It augments the experience, judgment, and skill that only a human can provide.

You, the construction professional, are the irreplaceable technology. Go learn the rest and make it work for you.

# Bibliography

It will surprise no one that I'm a big reader. All of these books, websites, and courses have had an influence on this book. I include a brief reason to read them where appropriate.

## Books

**Arthur, W. B.** (2009). *The Nature of Technology*. Free Press.
Will change your view of technology.

**Construction Specifications Institute** (2011). *The CSI Project Delivery Practice Guide*. John Wiley & Sons.

**Crawford, M.** (2009). *Shop Class as Soulcraft*. Penguin Books.
A love letter to what Crawford calls "the useful arts."

**Fischer, M., Ashcroft, H., Reed, D., and Khandoze, A.** (2017). *Integrating Project Delivery*. John Wiley & Sons.
Amazing book from some of the smartest thinkers in construction.

**LePatner, B.** (2007). *Broken Buildings, Busted Budgets*. University of Chicago Press.
Barry LePatner wrote *the* book on why construction contracts guarantee waste.

**Smith, D. K. and Tardif, M.** (2009). *Building Information Modeling*. John Wiley & Sons.
Considered one of the best books on BIM.

## Websites

**Construction Progress Coalition,** at https://www.constructionprogress .org/. Run by Nathan Wood and his team, they produce uniquely thought-through analyses of problems, most recently to do with RFIs.

**Dodge Analytics,** at www.construction.com. Some of the best information in the industry.

**JBKnowledge**, at https://jbknowledge.com/construction. James Benham and his team have been publishing great work on construction technology for longer than anyone.

**McKinsey Global Institute**, at https://www.mckinsey.com/industries/ capital-projects-and-infrastructure/our-insights. McKinsey generally views the world from the CEOs perspective.

## Courses

**Andrew Ng's AI courses**, at https://www.deeplearning.ai/. Founder of Google brain and of Coursera.org, Andrew's courses start math and code free, but you can dive deeper into harder levels if you want.

# About the Author

Hugh Seaton thinks technology is a useful tool for humans to get great things done, and has dedicated years to building products, helping others learn, and learning as much as he can along the way. The rest doesn't matter.

# Index

## A

accelerometers, 156
accountability, 3
accounting software, 69, 71
Adobe Photoshop, 40–41
adoption, product, 183–185
AEC Hackathon, 9
AEC (Architecture, Engineering and
    Construction) industry, 9
Agile Manifesto, 29
agile methodology, 29–30
AI (artificial intelligence). *See*
    artificial intelligence (AI)
Alexa, 107
algorithm, 102, 110, 111, 119–120,
    149
alpha testing, 36
AlphaGo, 112
Amazon, 35, 53, 125
analogy, thinking by, 176
analytics, data and, 59–61
Andreeson, Marc, 25, 34
Android operating system, 44
APIs (application programming
    interfaces). *See* application
    programming interfaces
    (APIs)
App Marketplace (Procore),
    54–55, 80

App store (Apple), 141–142
Apple
    App store, 141–142
    FaceID, 125–126
    iOS, 44
    iPads, x
    iPhone, 141
    operating system of, 43–44
    Siri, 36, 107
    technology of, 33–34
application programming interfaces
    (APIs)
    building around, 135–137
    limitations of, 64
    overview of, 53–55
    Procore and, 88–89
    products within, 56
    rise of, 77
    use of, 88–89
applications, 40–47, 52, 149
AR (augmented reality), 36–37, 92,
    148–153
AR cloud, 151–152
arcGIS, 55, 65
architects, 16–17, 95
Architecture, Engineering and
    Construction (AEC)
    industry, 9
artificial intelligence (AI)
    accuracy of, 128–132, 139
    Amazon and, 125